Calibrations

Modern human societies rely on measurements. Measurements are essential for understanding and describing the world around us and the overall nature. Ensuring measurement accuracy, traceability, and consistency is important in scientific research, trade, manufacturing, health and medicine, and the everyday running of modern societies. One of the most essential components of measurement is the calibration. This book walks the reader through fundamental principles of calibrations, traditional calibration methods, the related standards, and procedures, as well as the fast-developing and diverse range of artificial intelligence (AI)-based calibration methods.

Calibration complies with strict regulations and legislation by governments, relevant authorities, and concerned parties. New and effective calibration techniques are evolving in parallel to the traditional methods. As explained here, new techniques that are based on AI offer instantaneous, continuous calibrations in standalone devices as well as in complex and large systems. Continuous and large-scale calibrations are possible due to the new generation of computers and advanced digital systems, intelligent devices and sensors, communication networks, advanced mathematical methods, and the newly developed areas such as cyber-physical concepts.

This book highlights that advanced technological and AI methods are applied for the calibration of distributed cyber-physical and data-driven applications. AI-based calibrations find applications in large-scale systems, where traditional methods have become almost impossible to implement or can be costly. Typical examples are Internet of Things (IoTs), self-driving vehicles, intelligent transportation systems, unmanned aerial vehicles, drones and drone clusters, industrial processes, communication networks, health and medicine, environmental monitoring, pollution control, remote and inaccessible devices, smart cities, space explorations, management systems, economics, and finance.

Calibrations
Artificial Intelligence and Applications

Halit Eren

CRC Press
Taylor & Francis Group
Boca Raton London New York

CRC Press is an imprint of the
Taylor & Francis Group, an **informa** business

Designed cover image: Halit Eren

First edition published 2026
by CRC Press
2385 NW Executive Center Drive, Suite 320, Boca Raton FL 33431

and by CRC Press
4 Park Square, Milton Park, Abingdon, Oxon, OX14 4RN

CRC Press is an imprint of Taylor & Francis Group, LLC

© 2026 Halit Eren

ISBN: 9781032968070 (hbk)
ISBN: 9781032968087 (pbk)
ISBN: 9781003590767 (ebk)

DOI: 10.1201/9781003590767

Typeset in Times
by codeMantra

Contents

Preface

Modern human societies rely on measurements. Measurements are essential for understanding and describing the world and the overall nature around us. Measurement technologies enable a common understanding and communication of scientific and technological advancements. Ensuring measurement accuracy, traceability, and consistency is important in scientific research, trade, manufacturing, health and medicine, and daily running of modern societies. One of the most important components of measurement is the calibration.

Calibration is essential in engineering, science, and social science-based processes. Calibration is applied in many fields where measurement is involved for checking the performance of the process against expected values or known standards. The concept of calibration was developed over thousands of years for distance and weight measurements. Calibration became a methodical science during the Industrial Revolution, and developed significantly thereafter. In some cases, it may be sufficient to calibrate measurement devices against others, while in other cases, strict calibration processes must be followed, and verification documents must be obtained. After the calibration of a device or a process, the future operation is deemed to be error-bound for a given period under similar operating conditions.

Calibration methods for measurements are practiced in accordance with well-established rules and procedures. Calibration is covered by strict regulations and legislations that are implemented by relevant governments, authorities, and concerned parties. New and effective calibration techniques are evolving in parallel to the traditional methods. The new techniques are based on artificial intelligence (AI) and offer continuous calibrations in standalone devices as well as calibration of large systems. Continuous and large-scale calibrations are possible due to the new generation of computers and digital systems, intelligent devices and sensors, communication networks, advanced mathematical methods, and new application areas such as *cyber-physical* systems. Cyber-physical systems demand a new paradigm, a new way of thinking, and new technologies and techniques for successful calibrations.

Calibrations serve at least two purposes: (1) eliminate biases and minimize system errors and uncertainties, and (2) convert instrument readings to the units of interest if the instrument reads in different units. Calibration provides consistency in reading, addresses uncertainties, and reduces errors, thus validating the results of measurements universally. Without calibrations, process outcomes and expected results may be unreliable and misleading. In many cases, without calibration, it may be better not to proceed with measurement. Measurements may contain totally unrelated readings or may be totally unreliable, containing serious errors.

There are many measurements and calibration standards, and thousands of laboratories and institutions worldwide to implement those standards. Calibration involves special equipment, dedicated devices, appropriate software, relevant mathematical tools, and expertise. Engagement of these institutions requires a set of procedures and rules carried out by highly trained personnel conducting calibrations within strict rules and regulations. However, today, artificial intelligence techniques have made headway as alternative or complementary tools in calibrations. This book explains that there are many cyber-physical and data-driven systems that can only be calibrated by distributed computerized systems. Day by day, traditional methods are becoming less effective, as the modern operations diversify, vary, and become more widespread.

AI offers continuous and fast calibrations in many processes, devices, and clusters of devices. AI finds applications particularly in large-scale systems where traditional methods have become almost impossible to implement or costly. Typical examples are Internet of Things (IoTs), self-driving vehicles, intelligent transportation systems, unmanned aerial vehicles, drones and drone clusters, industrial processes, communications networks, health and medicine, remote and inaccessible devices, smart cities, space explorations, management systems, economics, and finance.

With the help of modern digital technology, many intelligent systems can achieve self-calibration in a continuous manner by using AI, lookup tables, charts, and curves, which make these systems autonomous, flexible, robust, and reliable. Implementation of AI-based calibrations is widely varied, ranging from clustering, regression, optimization, inference, and self-calibration. AI techniques naturally exhibit convenience in regression, clustering, optimization, and inference due to the inherent techniques they are built upon.

Dedicated to Emre, Alice, Violet, Leo, and Maxwell

Acknowledgments

I would like to express my thanks to Dr. Deniz Savas, Cahit Bückün, Numan Altekin, Prof. Lance Fung, Raymond Ryken-Rapp, Dr. Dincer Erer, and Dr Hilary Cussons for their encouragement and for continuous involvement in conversations with me during the development of this book. My gratitude is extended to my kids, Emre Eren and Pelin Eren, for patiently listening to me and providing information and guidance on some sections of this book. Finally, putting this book together would not be possible without the tireless involvement and encouragement from the personnel of CRC Press.

Disclaimer: I would like to confirm that no form of AI, such as ChatGPT and the like, has been consulted or used in producing the text in any part of this book.

Author's Biography

Halit Eren received a BEng, 1973, and an MEng, 1975, in electrical engineering and a PhD in control engineering, 1979, all from the University of Sheffield, UK. He obtained an MBA from Curtin University in 1999, majoring in international management. He worked in Etibank as a project engineer on automation and instrumentation in copper plants. He served as a lecturer at Hacettepe University-Beytepe Campus, 1980–1981, and as an assistant professor at the Middle East Technical University-Gaziantep Campus, 1982. He worked at Curtin University between 1983 and 2015, first at the School of Mines, Kargoorlie, and then at the School of Electrical Engineering, Bentley Campus, Perth, Western Australia. He served as an associate professor at Polytechnic University, Hong Kong, and as a visiting professor at the University of Wisconsin-Madison, USA. His academic duties involved research and teaching in control systems, digital transmission and interface design, process control and instrumentation, and engineering management. He has attracted numerous research grants from the government and industry, and has also acted as a consultant for many private and governmental organizations. He supervised and graduated PhD and master's degree students. He served as the head of the Department of Electronics and Electrical Engineering at Curtin University.

1 General Concepts

1.1 INTRODUCTION

Device and measuring instrument calibration is a critical component in maintaining the precision, efficiency, and safety of processes. Highest levels of accuracy, reliability, and certainty can be achieved by understanding the process, recognizing its importance, and implementing the best practices in many operations.

The first scientific papers appeared on calibration in 1888. With the advent of artificial intelligence (AI), the concepts in calibration have attracted renewed interest in recent years. Between 2020 and 2024, the number of papers that appeared in journals and conferences is about 25,000, which is more than 25% of the total papers since 1888. The increase in the number of calibration-related publications is significant because with the application of AI, many new and novel ideas are generated, a diverse range of methods are developed, new application areas are tried, and a high degree of reliability is achieved.

To understand and appreciate calibration, we need to understand some important terms often used to describe a particular aspect or the entire concept. Calibration and many related terms are defined and used with different meanings in various sources and disciplines. Even within the same discipline, the meanings and interpretations change due to long historical development in practices. Therefore, some terms may have different meanings, or different terms may describe the same practices.

Due to long historical roots, the vocabulary on calibration is very rich. In this chapter, some important terms are defined, and the need for calibrations are explained. The reasons of why calibrations are necessary and the ways calibrations are conducted are highlighted. Fundamental calibration methods are listed, and the sources of information on calibration are given.

All scientific and engineering calibrations, without exception, are based on SI units. Therefore, the fundamental measurement standards are explained, and paper standards, which are guidance for calibrations, are discussed.

1.2 DEFINITIONS

The terms often used in this book are discussed below in alphabetical order.

Oxford English Dictionary defines the term calibrate as "**1** mark (a gauge or instrument) with a standard scale of readings. **2** Compare the readings of (an instrument) with those of a standard. **3** Adjust (experimental results) to take external factors into account or to allow for comparison with other data.

Calibration in economics is defined as: "the process of adjusting a model's parameters so that its predictions match historical data." Statistical techniques are used to determine the model parameters (indicators) [e.g., prices, gross domestic product (GDP), consumption, investments, money supplies, international trade, etc.] for best fit with the practical data.

DOI: 10.1201/9781003590767-1

Calibration in finance is "the process of determining a parameter set that model prices and market prices match very closely for a given set of liquidity traded benchmark instruments." The calibration procedure delivers the optimal parameter set for the model based on calibration instruments. Depending on the application, some of the financial parameters are movements in profit and loss, balance sheet, cashflow prices, production levels, and exchange rates.

In measurements, the ISO International Vocabulary of Basic and General Terms in Metrology (VIM) defines calibration as[A1] "a set of operations to establish the relationship between values of quantities indicated by measuring devices or processes under specific test." Therefore, calibration assigns values to the response of some measurements to reference standards. It aims to eliminate or reduce bias in the system relative to the reference base in accordance with a specific algorithm. The bias may be caused by linear or nonlinear parameters drifting over the time of a process.

VIM goes further by clarifying calibration as: "operation that, under specified conditions, in a first step, establishes a relation between the quantity values with measurement uncertainties provided by measurement standards and corresponding indications with associated measurement uncertainties and, in a second step, uses this information to establish a relation for obtaining a measurement result from an indication."

Calibration is the comparison of measuring devices to a measurement standard of known accuracy. The purpose is to determine accuracy and uncertainties by adjusting control parameters.[A2] Therefore, calibration is a process of reporting measurement results by comparing them with a reference of verified accuracy. Accuracy and reliability of all measurements will be in doubt and untrustworthy if the instruments used are not calibrated. It is important to emphasize that calibration is not an adjustment for maintenance or repair of a device.[A3]

International Standards Association in Automation, Systems, and Instrumentation Dictionary defines calibration as "a test during which known values of measurand are applied to the transducer and corresponding output readings are recorded under specified conditions." The definition includes the capability to adjust the devices to zero and to set the desired span.[A4]

As can be seen in calibration, one or multiple variables are involved. These variables are used to calibrate the dependent variables of a particular process, albeit in science, engineering, economics, finance, social studies, and measurements. The aim of calibration is to fit a calibration function that describes the relationship between the variables of interest and make necessary adjustments for the best fit.[A5]

Accuracy is the quality or state of being correct. It is the closeness of the results of measurements or specifications conforming to the correct values of measurements and to the known standards. The more influential factors are considered in measurements, the more the likelihood of reaching better accuracy. Accuracy can be expressed in percent reading, measurement units, or percent of span.

Adjustment is "the set of operations carried out on a measuring system so that it provides prescribed indications corresponding to given values of a quantity to be measured." A series of coordinated adjustments may be necessary during the calibration of a measuring system.

Bias is the inclination or prejudice for or against one person or group. For example, bias in AI is the systematic errors or unfairness in an AI system. Bias leads to potential discriminatory or inaccurate outcomes. Sources of biases in AI originate mostly during the collection and selection of data, training, physical measurement bias, algorithmic and, human bias, and so on.

Device is a piece of electronic or mechanical equipment designed to serve a specific purpose or to perform a specific function. Devices include measuring instruments, sensors, data loggers, monitoring devices, analytical devices, and other devices that may be integrated into equipment or a process.

Inference is a conclusion based on evidence and reasoning. It has similar meanings to deduction, entailment, implication, or conclusion. Inference is an educated guess based on what is already known. In AI applications, "inference" is the process where a trained model uses the learned knowledge (trained knowledge) to draw conclusions or to make predictions from related unseen data.

Intelligent sensors are smart sensors that collect and analyze data and act upon it. Some intelligent sensors use AI-based internal algorithms to interpret past and present data and make their own decisions within the requirements of the system they are operating in.

Measurand is the quantity, object, or property that is measured. Measurand is the measured quantity such as volume, pressure, biopotential, temperature, voltage, height, distance, chemical concentration, displacement, velocity, flow, speed, weight, force, acceleration, light intensity, radiation, energy, phase shift, albumin, enzyme activity, and so on. Measurand specifies values of variable(s) relevant to the measured entity.

Multimodal has several different modes of occurrence. Multimodal is used often in measurements, calibrations, and AI applications with slightly different meanings. For example, multimode in mathematics is a set of data with four or more modes. Multimodal in statistics is a curve with several maxima. Multimodal fusion is a process of combining data from multiple different types of sensors and devices to achieve a comprehensive understanding of the performance of a system.

Performance check is the routine checking of the performance of a device to verify that it has remained within the specified range of accuracy, uncertainty, and precision.

Precision is the condition of being exact and accurate. In measurements, precision is the closeness of agreement between the results of successive measurements with the defined procedures under prescribed conditions.

Process is a series of actions or steps taken to achieve a particular result or a goal or a product. Some examples are *manufacturing processes* that involve multiple stages of transformation, testing, and quality control of materials. A *social process,* on the other hand, is the way living beings establish interactions and relationships with the environment in the context of cooperation, evolutionary changes, conflict, and competition. A *financial process* involves planning, execution, and monitoring by means of data collection and appropriate management.

Regression is a statistical method used to determine the relationship between a dependent variable (output variable) and one or more corresponding or influencing independent variables (input variables). It predicts and determines the strength and character of how one variable changes based on the other(s). Regression has different meanings in social science, for example, regression is a shift toward a less perfect state, or loss of acquired skills.

Replication is the action of copying or reproducing something. It is the repetition of a procedure or an experiment or a trial for the aim of obtaining consistent results. In calibrations, replication is important for the statistical determination of results to reduce uncertainty and improve accuracy. Random fluctuations can lead to small but substantial errors in measurements. By performing replications at each measurement, some or most of the errors due to random fluctuations can be identified, analyzed, and compensated for.

Reproducibility of a measurement is the closeness of agreement between results of measurements of the same measurand carried out under changed conditions. It is the ability to be reproduced or copied with consistency. In measurements, reproducibility is the degree of consistent results when the experiment or process is repeated.

Repeatability error is the deviation from the ideal or standard value when measurements are repeated. It indicates that when a procedure is repeated under identical conditions with the same inputs at different time intervals, the response (output) may not be the same as the previously recorded values.

Self-adjustment is the process of self-removing systematic errors [Metrology and the International Vocabulary of Basic and General Terms in Metrology (ISO VIM)]. [A1] It is the ability of a device or a system or a process to make changes within itself or adapt to its environment without external intervention. In calibrations, options for self-adjustment include hardware adjustments (manual or automatic), AI models, calibration algorithms, and stochastic and statistical mathematical tools.

Smart sensors can sense variables, process, and analyze the information attained and communicate the data using various communication methods and networks. Smart sensors use microprocessors and other electronic components to perform the required functions such as conditioning, amplification, filtering, conversion of signals, collecting and storing data, self-monitoring, self-adjustment, and self-calibration.

Tolerance is the permissible deviation from a specified value. Tolerance can be expressed in measurement units, percent of span, or percent of reading.

Uncertainty is the state of being uncertain. In measurements, uncertainty is the statistical dispersion of the values that can reasonably be attributed to a quantity measured on an interval scale. It is the margin and significance of doubt that exists in the results of measurements.

Unimodal has one mode. In statistics, unimodal is a symmetric or skewed probability distribution that has one peak. An example is the normal distribution (Gaussian distribution). Other distributions can be uniform, bimodal, or bell-shaped.

Validity of calibration is a range of values of all significant influence quantities for which the calibration results are valid. The validity conditions also include the number of measurements used to compute a result and the methods used in the analysis of results, since repeated measurements by an instrument may yield different results.

Verification is the process of establishing the truth, accuracy, or validity of something. It gives objective evidence that a given process fulfils specified requirements.

1.3 WHAT TO CALIBRATE

Any dependent variable of importance requires calibration. Calibration of devices, sensors, and instruments is essential to assess the validity and relevance of related measurements in engineering, science, research and design, medicine, health, and many other applications. Calibrations can be carried out on (1) a single device working independently, (2) multiple devices working independently but part of a system, (3) multiple devices operating cooperatively, (4) multivariate devices, and (5) clusters of devices made up of many independent but related measurements.

Before the emergence of cyber-physical systems, calibrations were based mainly on the calibration of single devices or small groups of devices. But today, there are many operations that involve thousands, even millions of devices working in unison, thus requiring new calibration techniques. Most of the new calibration techniques involve sophisticated algorithms and a diverse range of AI models. AI models have a broader range of calibration applications in social sciences, finance, economics, education, politics, and so on. As in the case of science and engineering applications, calibrations in social sciences require appropriate measurement methods, data gathering, data processing, and analysis.

The selection of relevant calibration methods depends on the physical and operational features of processes or devices under test. A few introductory examples of what to calibrate are given below.

Flow calibration: There are many different forms of physical flow in gases, fluids, air, and solids. Therefore, many different measurements methods evolved over decades using suitable measuring techniques and devices. For example, in the case of static-gravimetric liquid flow, a calibration facility may include a reservoir, a pumping system, a pipeline, a flowmeter located on the pipeline, data collection system, computers and interface, supporting software backed up with relevant mathematical tools, and AI models.

Sensor calibration: Today's complex and intelligent systems are highly dependent on the correct operations of the sensors. Sensors are calibrated often in static form before they are integrated into a system and in dynamic form after the integration. Static and offline calibrations of sensors can be done by injecting time-varying input signals in laboratories. Dynamic calibration becomes necessary once a sensor is integrated into the systems. The performance of a sensor can drift substantially and encounter noise and other unaccounted influences. Unless recalibrated, additional influences can introduce inaccuracy and uncertainty or even resulting in total failure.

Calibration of food products: Since food contains many different forms of chemical substances, calibration of devices in the food industry can be complex. As an example, let's take the case of "honey." Parameters such as fructose, glucose, furanose, maltose, water content, and acidity need to be identified and calibrated separately, as well as the need for an aggregate calibrated result.

Calibration of images: Calibration is one of the first steps in image processing. For example, images in space explorations are calibrated to eliminate the effects of cameras, light pollution, and distortions. Various methods are used to ensure the quality of calibrated images by eliminating thermal, readout, and other effects. For thermal effects, the cameras may need to be cooled below certain temperatures, dark frames can be used to compensate for the noise generated by camera electronics, and so on.

1.4   WHY CALIBRATE

Calibration is a process of testing and comparing the errors with expected outcomes or agreements with related standards. Calibration assures that a device or a process meets expected performance specifications within ubiquitously acceptable levels of certainty and accuracy.

There are many technical and legal reasons for performing calibration. The brief list below highlights some of the reasons why calibration is essential:

1. Establishing and demonstrating accuracy
2. Ensuring consistency and traceability to international standards
3. Determining uncertainty budgets
4. Assessing reliability, repeatability, and reproducibility
5. Addressing legal and trading requirements
6. Errors can accumulate and propagate in large and massive-scale systems
7. Administrative, procedural, and documentary requirements
8. Accreditation and quality assurance needs
9. Public safety and operational safety
10. Downstream process efficiency
11. Assessing the health and validity of operations before calibration
12. Assuring consistency and compatibility with those devices calibrated elsewhere
13. Reducing legal liability and mitigation against legal challenges
14. Maintaining the quality of products
15. Addressing the requirements of standards such as ISO 9000, ISO 1400, and QS-9000
16. Life analysis of devices, determining depreciation and replacement needs
17. Promoting global acceptance of products to gain advantage in competitiveness

As technology changes, the regulations and legislation change continually, and calibration helps compliance with these changes.

1.5   HOW TO CALIBRATE

Calibration requires adjustment and setting of parameters, which can be done in three basic ways.

1. Hardware calibration by adjusting parameters and settings
2. Software calibration by adjusting parameters and settings
3. Combination of hardware and software calibration

Hardware calibration involves the adjustment of relevant controlling components in mechanical systems, electronic circuits, chemical, or biological processes.

In electronic measurement applications, the adjustable components in circuits can be digital processors, analog to digital converter and digital to analog converters, filters, amplifiers, voltage-controlled oscillators, reference voltages, multiplexers, power supplies, digital pods, bias circuits, resistor banks, capacitor banks, and many others. These components are designed and manufactured for adjusting manually or by using suitable software. Hardware calibration processes need to comply with ISO standards.

Software calibration is largely based on historical data and ground truth information to mitigate device errors. Software calibrations use statistical tools and offer many new and unconventional calibration methods including AI models.

Hardware and software calibration is based on adjusting some of the hardware components and altering some parameters in software to mitigate errors.

1.6 DIAGNOSIS FOR CALIBRATION

Diagnosis of errors involves observation of variations in the recorded outputs of a calibrated device. These outputs need to be compared with the expected values or known standards. Variations can be detected by drift detection algorithms, continuous monitoring, AI methods, or human observation. The sources of errors and variations can be due to sensor drifts, malfunctioning of supporting electronic components, nonlinearities in responses, aging, hysteresis effects, ambient noise, external interference, and prevailing environmental conditions of use.

The variations are detected by observing zero shift, span shift, linearity, or hysteresis errors. Anomaly detections can be used by applying single test or multipoint tests on the operation of the device or the process. Malfunctioning and failure are indications of the need for calibrations or need for repairs or replacement of measuring devices.

1.7 LITERATURE ON CALIBRATION

Information on calibration is available in a wide range of sources (Figure 1.1), some of which are given as follows:

1. *Scientifically published papers* provide a wealth of information on calibrations. Scientific publications on calibrations started in the late 1800s in the IEEE Xplore. There are over 93,000 publications containing the word "calibration", and it is still growing.
2. *Manufacturers* supply comprehensive information on calibration requirements and procedures for the equipment they manufacture.
3. *Regulation authorities and standards institutions* supply information on the calibration of sensors, instruments, equipment, and devices. Calibration can be a statutory requirement in many measurement applications such as health and safety.
4. *Calibration services* provide a wealth of information on procedures and practices.
5. *Organizations* provide rules and regulations on calibration and assurance planning.
6. *Books* such as this one provide comprehensive information on calibrations.

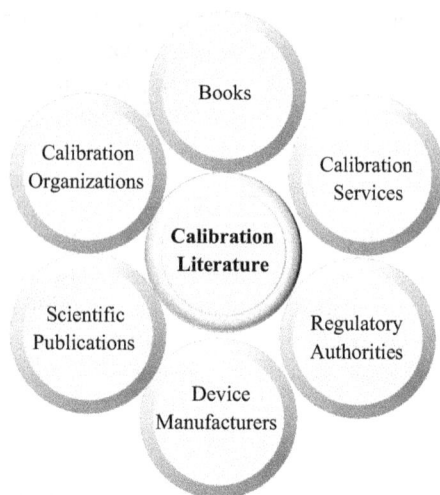

FIGURE 1.1   Sources of information on calibration. Calibration is essential in all types of measurements. Therefore, there is a wealth of information that can be obtained from different sources. Many instrumentation- and measurement-related books contain chapters or sections on calibrations. Other sources are books dedicated purely to the calibration concepts such as this one. Manufacturers' recommendations, calibration and standardization organizations, and scientific publications are typically valuable sources.

Calibration is a highly established science. There are many accreditation bodies worldwide. Numerous nations and organizations maintain laboratories, with one of their primary functions being the calibration of measuring devices. A good example of such an organization is the International Laboratory Accreditation Cooperation (ILAC).[A6] This organization was formed in the year 2000, having only 28 members. Today, ILAC has 152 members from 128 different countries. ILAC is supported by 103 full signatory members, 15 associates, 6 affiliates, 22 stakeholders, and 6 regional cooperation bodies. ILAC covers testing, calibration, inspection, proficiency testing of providers, and reference material producers. Today, it houses over 76,000 accredited laboratories, 10,500 accredited inspection bodies, 400 accredited proficiency testing providers, and over 30 reference material producers. In addition to calibration laboratories, ILAC houses testing laboratories (ISO/IEC 17025), medical testing laboratories (ISO 15189), inspection bodies (ISO/IEC 17020), proficiency testing providers (ISO/IEC 17043), and reference material producers (ISO 17034). Some examples of member organizations are the Standards Council of Canada (SCC), National Institute of Standards and Technology (NIST) of the USA, National Association of Testing Authorities (NATA) of Australia, NAAU National Accreditation Agency of Ukraine, British Calibration Services (BCS), and many others.

1.8   MEASUREMENT STANDARDS

The Oxford English dictionary defines *standard* as a **noun:** (1) a level of quality or attainment, (2) a required or agreed level of quality or attainment, and (3) something used as a measure, norm, or model in comparative evaluations. **Adjective:** (1) used or

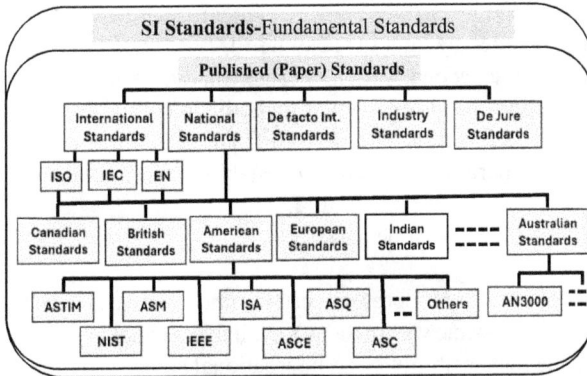

FIGURE 1.2 A comprehensive guide to measurement standards. Standards provide essential guidance on devices, instruments, instrumentation, measurements, and processes. All calibrations can be traced back to SI units or units derived from them. Even if different units are used, such as pound, feet, and yards, they can all be converted to equivalent expressions to SI units. Complying with the SI units is a must, and the paper (published) standards provide guidance on calibrations.

accepted as normal or average, and (2) regularly used or produced such as of a size and measure.

Measurement standards are classified as primary standards and secondary standards. Primary standards are the fundamental physical standards accepted internationally. Secondary standards are derived from primary standards and kept by nations and organizations.

Standards in measurements are comparative evaluations of physical quantities, processes, and procedures. Standards are available (Figure 1.2) as:

- Fundamental physical standards,
- Paper (published) standards published by national and international standardization organizations. They are available for anyone to use,
- De facto standards that are followed by organizations due to convention or popular use, and
- De Jure standards that are part of legally binding contracts, laws, or regulations.

Fundamental (primary) physical standards are internationally accepted standards. Primary standards are "meter" for length, "second" for time, "ampere" for electric current, "Kelvin" for temperature, "kilogram" for mass, "mole" for the amount of substance in chemical science, and "candela" for luminous intensity. SI sets defining constants to determine the fundamental standards. These constants are the cesium hyperfine splitting frequency, the speed of light in vacuum, the Planck's constant, the elementary charge on a proton, the Boltzmann constant, the Avogadro constant, and the luminous efficacy of a specified monochromatic source.

The primary standards of physical entities for length and weight (the meter and the kilogram) are kept in the International Bureau of Weights and Measures in Sèvres,

France. Apart from the physically existing standard, the meter was defined as the length of the path traveled by light in vacuum in 1/299,792,458 of a second. The standard unit of time is the second, which is established in terms of known oscillation frequencies of certain substances, such as the radiation of the cesium-133 atom. The standards of electrical quantities are derived from mechanical units of force, mass, length, and time. Temperature standards are established as international scale by taking 11 primary fixed points of temperature.

Standards are observed in three stages:

1. International standards represent certain units of measurement with maximum accuracy possible within today's available technology. These standards are under the responsibility of an international advisory committee and are not available to ordinary users for comparison or calibration purposes.
2. Secondary standards are independently calibrated against the international standards by absolute measurements. Secondary standards are maintained in national laboratories and by some industrial organizations. They are periodically checked against primary standards and certified.
3. Working standards are used to calibrate general laboratory and field instruments.

Published standards (also known as Paper Standards) are documents that provide textual and illustrative information on what and how the processes and practices should be performed to achieve a particular task. The National Standards Policy Advisory Committee of Australia[A7] describes paper standards as "prescribed set of rules, conditions, or requirements concerning definitions of terms: classification of components; specification of materials, performance, or operations; delineation of procedures; or measurement of quantity and quality in describing materials, products, systems, services, or practices."

Published standards are necessary for worldwide trading, business practices, production of goods, manufacturing measurement devices, scientific investigations, and research and development. They ensure consistency in the implementation and interpenetration of different technologies in different organizations and sectors. Because of the national and international implications, there are many institutions that are responsible for developing, investigating, and maintaining the relevant standards to support worldwide scientific, commercial, and industrial activities.

De facto standards are practices that are accepted and used widely, but they have not been formally sanctioned by standard authorities. De facto standards emerge by market dominance and widespread use. Typical examples of de facto standards are the HTTP protocol for communication between web browsers and servers, typewriter and computer keyboards.

De Jure standards are recognized by law. They are endorsed by formal standard organizations like ISO, ANSI, IEEE, and others. They can be part of legally binding contracts or legislation or regulations. Some examples of de jure standards are the ASCII character set, internet TCP/IP, Code Division Multiple Access (CDMA), and so on.

FIGURE 1.3 Traceability pyramid. The traceability pyramid illustrates how the compliance-calibrated devices are traced back to the SI standards. International, national, and company standards are the next tiers to propagate the primary standards to the expectations in calibrations to the level of working and device-level operations.

In calibrations, the traceability to international standards needs to be maintained for all calibrated devices. This is illustrated in the traceability pyramid in Figure 1.3.

In summary, in measurement and calibration literature, standards are applied as follows:

- International standards
- National standards
- Organizational standards
- Paper standards
- Intrinsic standards
- Reference standards
- Master standards
- Working standards
- Derived standards
- Consensus standards
- Transfer standards

International standards are the most recognized standards for calibration. International standards are the highest level of reference standards agreed by multiple countries for the common purpose (kept at the Bureau of Weights and Measures in Sevres, France),

At the international level, decisions concerning the International System of Units (SI) and the realization of the primary standards are taken by the Conférence Générale des Poids et Mesures (CGPM).[A8] The development and maintenance of primary standards is coordinated by the Bureau International des Poids et Mesures (BIPM), which also organizes intercomparisons at the highest level. The ILAC promotes laboratory accreditation and the recognition of competent calibration and test facilities around the world.

National standards are derived from the primary standards. The "relationship" between measured or indicated values and those of the reference values is a key issue with regard to calibration. The calibration process includes a wide variety of activities, including determining the mathematical measured values and the reference values. In the USA, national standards are maintained by NIST.

Organizational standards are standards kept by organizations that comply with national and international standards. The reference standard is "Measurement standard having the highest metrological quality available in an organization."

Paper standards, layout procedures, methods, and rules are explained above. Some of the published standards are accepted internationally; an example is the ISO 17025. ISO 17025 requires labs to demonstrate that they practice good calibration procedures backed up with well-defined methods, up-to-date equipment, trained personnel, well-maintained documentation, and quality management. Information on ISO 17025 can be found in many books and ISO publications.[A9]

Intrinsic standards are standards based on laws of physics, fundamentals of nature, and invariant properties of materials.

Reference standards are the highest metrological quality located at a site where calibration is being conducted.

Master standards are a lower level of reference standards and are used for lower-level calibration requirements of measuring devices.

Working standards are compared with master standards or reference standards on a regular basis. They are used for daily checking and comparison of the measuring devices.

Derived standards are a combination of two or more standards for the purpose of fulfilling traceability requirements.

Consensus standards are used when no traceability to a known standard can be established, but by an agreement of all parties observing the standard; an example is the Rockwell hardness testing process.

Transfer standards are applicable for transferring a device designed to operate in one location under set conditions but being transferred to another location.

1.9   CONCLUSIONS

A general outlook on the calibrations is presented. Some of the important but inexhaustive terms are defined. Sources of information on calibrations are given. Essential calibration methods are explained. The fundamental international measurement standards are discussed. The importance of paper standards is discussed in detail. It has been shown that calibration is an important process for the well-being of society.

REFERENCES

A1. JCGM 200:2012 International Vocabulary of Metrology – Basic and General Concepts and Associated Terms, 2012. https://doi.org/10.59161/JCGM200-2012, 2012 (Accessed on 18 April 2025).

A2. M. S. El-Azazy, "Analytical Calibrations: Schemes, Manuals, and Metrological Deliberations," *Calibration and Validation of Analytical Methods – A Sampling of Current Approaches*," InTech, London, UK, pp. 17–34, Ch. 2, https://cdn.intechopen.com/pdfs/58230.pdf (Accessed on 25 April 2025).

A3. ISA Instrumentation, *The Automation, Systems, and Instrumentation Dictionary* (4th ed.). Research Triangle Park, NC: ISA-The Instrumentation, Systems, and Automation Society, 2003.

A4. WHO Library Cataloguing-in-Publication, "Laboratory Quality Standards and Their Implementation," Jan. 2011. https://www.who.int/publications/i/item/9789290223979 (Accessed on 18 April 2025).

A5. A. Eroglu and M. N. Mahmoud, "Artificial Intelligence Based High Power Calibration Method for RF Pulse Amplifiers," *2024 IEEE 42nd VLSI Test Symposium (VTS)*, Tempe, AZ, USA, pp. 1–5, 2024.

A6. Home – ILAC International Laboratory Accreditation Cooperation. https://ilac.org (Accessed on 18 April 2025).

A7. National Standard Advisory Committee – DAFF. https://www.agriculture.gov.au (Accessed on 18 April 2025).

A8. *General Conference on Weights and Measures*, 27th meeting-15–18 November 2022. https://www.bipm.org/en/cgpm-2022 (Accessed on 18 April 2025).

A9. ISO/IEC 17025:2017 – General Requirements for the Competence of Testing and Calibration Laboratories https://www.iso.org/standard/66912.html (Accessed on 18 April 2025).

2 Calibration Strategies and Techniques

2.1 INTRODUCTION

Calibration can be classified as primary calibration and secondary calibration. The *primary calibration* is conducted against the primary standards. Secondary calibrations are used for calibrations of other devices with less expected accuracy. The calibration process transfers a reference value, such as the International System (SI) units, to the measuring device under calibration.

Secondary calibration can further be divided into two groups: (1) direct calibration and (2) indirect calibration. In direct calibrations, the device under test is compared with a standard device of known accuracy. In indirect calibration, the equivalence of two different devices is assumed to be similar.

The calibration process includes a wide variety of activities, including the determination of mathematical relationships between influencing factors and their effects on device output. The parameters are adjusted to correct the effects of known systematic factors to achieve good accuracy and reduce uncertainty.

Calibration in applications faces computational and statistical difficulties. Computational difficulties are related to time-demanding modeling and relevant algorithms for optimization and inference. There is a trade-off between accuracy and computation time. Statistical difficulties arise from the incompleteness of data and model representation and the existence of multiple possible solutions. Determination of uncertainty from the observations, models, and theoretical solutions may pose extra problems.

A rich volume of strategies and techniques has been developed after having practiced calibrations for many decades. This is due to the use of a wide variety of devices and their diverse operational requirements. This book visits some of the well-established calibration techniques, listed below:

1. Adaptive and continuous calibration
2. AI-based calibration
3. Automatic calibration
4. Change- and anomaly-based calibration
5. Cyber-physical calibration
6. Data-driven calibration
7. Database calibration
8. Dynamic calibration
9. End-to-end calibration
10. e-calibrations using AI
11. Field calibration
12. Hardware-based calibration

DOI: 10.1201/9781003590767-2

13. Intelligent calibration
14. Internet (i) -and e-calibrations
15. Laboratory-based calibration
16. Large-scale and massive-scale calibration
17. Low-cost sensor calibration
18. Model-driven calibration
19. Multidevice calibration
20. Multifunction calibration
21. Multivariate calibration
22. One point, two points, and multipoint calibration
23. On-the-fly and in situ calibration
24. Remote calibration
25. Self-calibration
26. Sensor calibration
27. Signal-based calibration
28. Single device calibration
29. Social science calibration
30. Software-based calibration
31. Static and dynamic calibration
32. Statistical and stochastic calibration

Detailed discussions can be introduced on all these techniques; due to a lack of space, essential and underlying principles will be provided in the following sections in alphabetical order.

2.2 ADAPTIVE AND CONTINUOUS CALIBRATIONS

The word *adapt* means adjusting or making something for a new purpose, or making it suitable for new situations. It also indicates becoming adjusted by self-adjustment or being induced to new situations. Adaptive calibration can be achieved through external adjustments or self-calibrations.

Continuous calibration is the calibration of a device or a system continuously. AI models and intelligent devices ensure that the system always remains optimally calibrated. Continuous calibration does not need human intervention but requires machine-to-machine communications. It uses algorithms to track the operational parameters continuously and calibrate any deviations from the expected ones. AI models are extensively used in continuous calibrations, thus opening new frontiers.

The application of adaptive and continuous calibration is diverse, ranging from industrial operations to biological, social, and psychological adaptations. Adaptive calibration is used in communication networks, automatic control systems, robotics, robot hand-eye coordination, and so on. Good examples of devices that adaptively calibrate themselves are used by disabled persons. A few examples are adaptive computers, adaptive brails, adaptive music controllers, adaptive hearing aids, adaptive wheelchairs, and so on.

In industrial automation and control systems, adaptive calibration aims to alter and optimize operating conditions of equipment, measuring devices, and sensors. Adaptive calibration is to capture and make changes in the operational characteristics of the sensors or soft sensors.

Adaptive and continuous calibrations, in social sciences such as economics, can imply the adjustment of a broad range of model specifications with microeconomics data and/ or macroeconomic data for developing alternatives for changes in real time. In biological systems, adaptive calibration models postulate individual differences in stress responsiveness and the ability of organisms to modify their developmental trajectories to match the local environmental conditions, and to the social and physical surroundings.[B1]

AI models are effective and suitable for adaptive calibrations since they are based on continuous learning and the capable of adapting to changes that take place over time. AI models are applied in adaptive calibration in processes or devices by considering unique characteristics of the process, environmental factors, and data from historical performances.

2.3   ANN AND ARTIFICIAL INTELLIGENCE-BASED CALIBRATIONS

The use of AI technologies is causing a paradigm shift in the way calibrations are performed, perceived, and used in a broad range of disciplines. AI models enable continuous calibrations in individual devices as well as large systems such as IoTs. AI models use calibration data for regression, clustering, and decision making in many real-time applications. AI helps with the automation of repetitive and routine tasks such as data entry, report generation, checking for standard compliance, and so forth.

AI-backed real-time calibration capabilities ensure continuous adherence to international standards. In the automated calibration systems, AI offers a new era of determining accuracy and uncertainty, enabling self-adjustments, and diagnosing errors. In many applications, the errors can be anticipated with a high degree of precision, efficiency, and reliability even before they occur. In industrial settings, an advantage of AI-driven calibration is that processes can continuously be monitored, ensuring compliance with international standards and regulations.

Evaluating the performance of a device involves consideration of various crucial factors such as accuracy and sensitivity. Accuracy directly impacts the reliability of the parameters derived from the measurements. Conversely, sensitivity determines the smallest detectable change in readings, accounting for its internal noise. AI-based calibration techniques are designed to enhance the accuracy and reliability of the data collected. Calibration techniques can be categorized into two distinct parts: *external calibration* and *internal calibration*. External calibration involves using external parameters or references or targets to calibrate. Internal calibration is achieved using device parameters. The utilization of the AI method shows promising potential in reducing the need for frequent internal calibration.

When remote and massive scale calibrations are involved, the use of wireless communication technology and remote monitoring techniques backed up by AI models enhances the effectiveness and efficiency of the calibration processes.

AI models in calibration can identify patterns in large data sets, analyze, and compare historical calibration data, predict equipment performance, and anticipate when a device requires recalibration by making real-time decisions.[B2]

Artificial neural networks (ANNs) are extensively used in calibrations because methods like feedforward networks are universal approximators capable of learning continuous functions with any desired degree of accuracy. In most cases, the ANN models are trained using the backpropagation (BP) algorithms. With high self-learning capabilities, ANNs offer a reliable tool to predict complex problems and capture complex interactions between different variables. They are suitable for problems that are complex to solve by conventional mathematical modeling or statistical descriptions. Neural networks are extensively used for linearization, in which the transfer function response curve of the sensor can be identified.

As in the case of many applications, ANNs in calibrations require careful determination of the number of neurons, number of layers, activation functions, training algorithms, and other computational requirements. Usually, the design of the network architecture starts with fewer hidden neurons and then an increasing number of hidden neurons until satisfactory results are achieved. Multiple layers of neurons with nonlinear transfer functions allow the network to learn the linear and nonlinear relationships between the input and output variables.[B3]

2.4 AUTOMATIC CALIBRATION

Automatic calibration (auto-cal) is the self-calibration of devices by comparing current data from measurements with the stored data. Auto-calibration determines equivalent parameters of equal quality in a process or a device. In many applications, optimization methods are used, which utilize similarity measures to quantify the differences between device data and historic data or data obtained elsewhere, such as the simulation results.[B4]

Many modern measuring devices offer features for automatic calibration so that calibration can be carried out without tempering the components. Once the calibrating device is connected to the computer, the appropriate software generates the necessary calibration information. Errors due to gains and offsets of the instrument are corrected mathematically by the onboard software to obtain the desired measuring values. Analog corrections can also be made via the adjustment of electronic components such as the analog-to digital and/or digital-to-analog converters, variable resistors, adjustable capacitors, and so on. Auto-calibration finds wide applications in image calibrations, mobile robots, and environmental applications, where multisensory modalities are used.[B5]

AI-based auto-cal is becoming common in multidevice and multisensory calibrations. The collected calibration data are shared and compared, and the parameters are adjusted for the complete system performance. Methods such as random forest regression (RFR), support vector regression (SVR), 1-D convolutional neural network, and 1D-CNN long short-term memory network models are some of the AI methods used in automatic calibrations.[B6]

2.5 CHANGE- AND ANOMALY-BASED CALIBRATION

Change detection (CD) is a method of monitoring unusual performances in measuring devices and processes. Unexpected changes, shown in Figure 2.1, are good indications of a device misreading or failures. Change and anomaly detection find applications in environmental monitoring, industrial operations, smart cities, home security systems, and finance. Change detections are used in calibrations in univariate and multivariate data streaming or batch-wise operations. There are many change

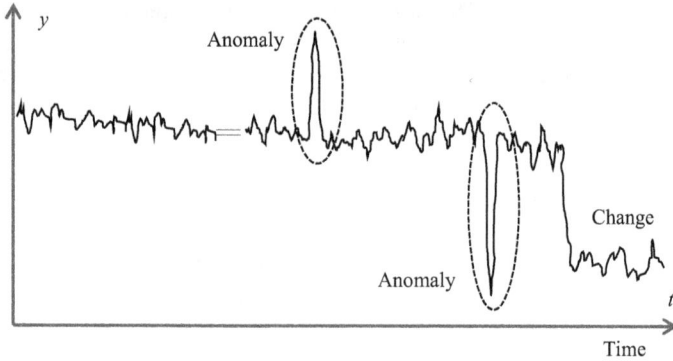

FIGURE 2.1 Change and anomaly behavior in measurement. Device operations are not always consistent and uniform due to internal and external factors. Devices can exhibit anomalies in readings by unexpected sudden changes but recover later to resume normal operations. The readings change unduly; it may be an indication of device failure, need for calibration, or unaccounted changes in operational conditions.

detection algorithms such as the MultiModal QuantTree, which detects changes by means of modality-specific statistics.[B7] Typical AI methods are the decision trees and DeepLog, both of which are toolkits for log analysis in automated anomaly detections.[B8]

Anomaly detection is an important method that has been studied extensively within diverse research areas and in various applications. Real-world datasets are analyzed to determine outliers and instances that stand out as dissimilar to all others. The common causes of outliers and anomalies in a data set are human errors (e.g., wrong data entry), measurement errors, data extraction errors, experiment planning/execution errors, intentional, intentionally introduced dummy errors and outliers to test detection methods, data processing errors, sampling errors, mixing data from faulty sources, and the natural errors inherent in the data.[B9]

Various AI models are developed for anomaly detection for supervised, unsupervised, and hybrid models. Anomaly detection finds applications in fraud detection, malware detection, intrusion detection, medical anomaly detection, log anomaly detection, Internet of Things (IoT), industrial systems, video surveillance, and many others.

2.6 CYBER-PHYSICAL SYSTEM (CPS) CALIBRATIONS

CPSs are defined as systems where the physical components are deeply intertwined with the software components to function in unity. Such systems exhibit distinct behavioral patterns and require different calibration procedures due to the cooperative and inseparable operation of the physical system and the software. Typical applications are in green transportation and electric vehicles. In such systems, in addition to physical resources, communications and control compose the cyber part of the system, thus adding extra complexities in calibrations. Components of a CPS are

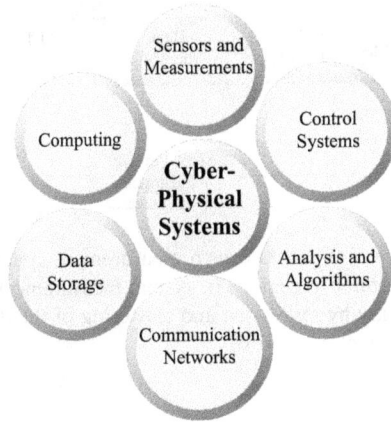

FIGURE 2.2 Components of a cyber-physical system. Cyber-physical systems are gaining a wide range of usage in many novel applications. They are complex due to intensive use of digital techniques, communication networks, data handling, and supporting software. Calibration of such systems requires new and sophisticated techniques.

illustrated in Figure 2.2. Substantial nonlinearities and uncertainties may be introduced due to complex operations of cyber and physical resources.[B10]

2.7 DATA-DRIVEN CALIBRATIONS

Modern calibration methods heavily rely on the data obtained from the related devices. Once the calibration data are generated, they can be used for making corrections, setting accuracy, and calculating uncertainties. With the use of appropriate software, the collected data can be applied on similar device calibrations operating under similar conditions, as in the case of environmental monitoring applications.

In data-driven calibrations, a large amount of data are collected on a system online and/or offline. Offline requires prior data collection under static or dynamic operational conditions. Once the initial data are suitably processed, AI or other techniques such as statistical methods can be employed for calibrations (Figure 2.3). In the case of online calibrations, the AI models can predict parameter values based on unknown affecting variable or variables, which are difficult to integrate in normal calibration procedures.

Data collection for calibration involves three steps: measurement, estimation, and compensation. Various ANN and AI methods are applied in data-driven calibrations such as the Extreme Learning Machine, Back-Propagation Neural Network, and Radial Basis Function Neural Network.[B11]

2.8 DATABASE CALIBRATIONS

Database calibrations rely on the data generated by a device or multiple devices residing in databases such as cloud, edge devices, or local databases. Data obtained from calibrated devices are often kept in databases, so that they can be reused on other devices for adjustments and to validate operating in similar situations.[B12]

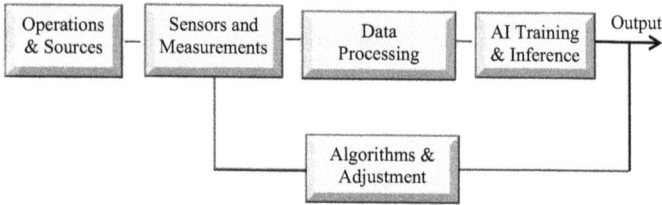

FIGURE 2.3 Block diagram of a data-driven measurement system. Data-driven systems involving AI methods determine variations in system measurements. The affecting factors can be identified conveniently by regression and clustering of the data to find out the basic reasons for variations over time.

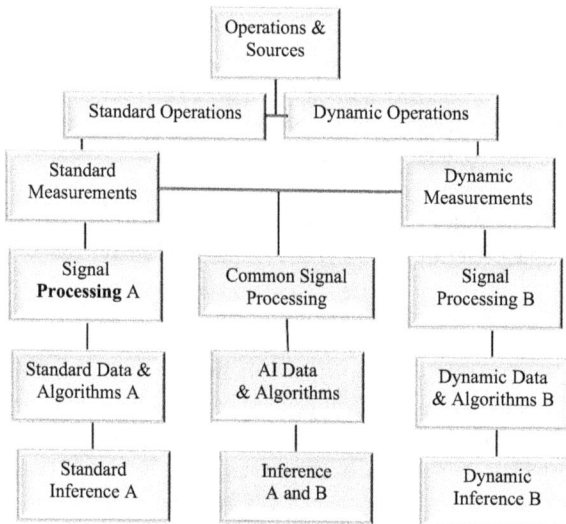

FIGURE 2.4 Block diagram of a dynamic measurement system. Dynamic systems can exhibit totally different measurement characteristics and parameters. In many dynamic systems, it may be necessary to configure the measurement system suitable for monitoring the parameters by using different sets of devices and signal processing techniques and algorithms. In other systems, common measurement devices and algorithms may be sufficient.

2.9   DYNAMIC CALIBRATIONS

Dynamic calibration has two meanings. The first meaning is when a device operates in transient conditions beyond its measurement range. The second meaning is when a device is integrated into a dynamic system performing its expected operations, which is explained in detail in Section 2.27. The procedure for calibrating dynamic and steady-state operations is illustrated in Figure 2.4.

Dynamic calibration is required when the condition of a device is significantly different from the steady-state operations. Under transient operating conditions, the normal parameters describing the operation can change substantially on a relatively shorter time scale. Examples requiring dynamic calibrations are shockwaves, explosions, unexpected pressures such as earthquakes, and blasts. In such operations, a comprehensive approach is necessary for mathematical modeling, which encapsulates the transient characteristics of the process. If the characteristics are significantly different, measurements require more suitable and appropriate sensors that can quickly switch from steady-state operations. The calibration process consists of a series of carefully worked out steps to ensure successful calibration. Performing these tasks entails applying computational methods of different types, both in a symbolic and numerical sense.[B13]

In many situations, in addition to the dynamic characteristics of the process, the dynamic response characteristics of the measurement devices must be evaluated independently from the steady-state values. The dynamic parameters of the measurand such as the amplitudes and frequencies are most likely to be altered significantly. The values of dynamic errors and uncertainty of the sensors are likely to be different compared to steady-state errors and uncertainties; hence, a new set of calibration methods is needed, as in the case of investigations of explosions and shock waves. The responses of measuring devices for each investigation may be very different in each case, even though processes are repeated under similar conditions.[B14]

Several methods have been used to address the problem of dynamic calibration. When physics-based models are used, the mathematical descriptions of dynamic systems may be based on known physical principles, making it easy for calibrations. For difficult or unattainable mathematical and statistical descriptions, AI methods are used to get estimates and inferences. Some examples of popular methods used in dynamic calibrations include AI models, iterative reweighted least squares schemes, Kalman filters (KF), extended KF, Unscented KF, particle filters, or Bayesian inference methods using Markov chain Monte Carlo.[B15]

In literature, it is possible to see the term dynamic calibration used for in situ calibrations or on-the-fly calibration.

2.10 END-TO-END CALIBRATION

End-to-end calibration is the complete calibration of a measuring system. Calibration starts (shown in Figure 2.5) from physical inputs, device outputs, front-end processing of signals, generation of data, managing the software, data transmission, and processing the data at the receiving end. The validity of calibration needs to be assessed at the end to verify correctness and to identify possible sources of errors.[B16]

End-to-end calibrations find wide applications in remote calibrations, industrial applications, and in many systems involving multi-modal (multiple different types) sensor fusion. Devices are calibrated intrinsically relative to their internal operations and extrinsically with respect to the other devices. Joint processing and optimization of intrinsic and extrinsic end-to-end calibrations improve accuracy, leading to better performance of the system compared to isolated calibrations of individual devices. There are many applications of AI models for end-to-end calibrations.[B17]

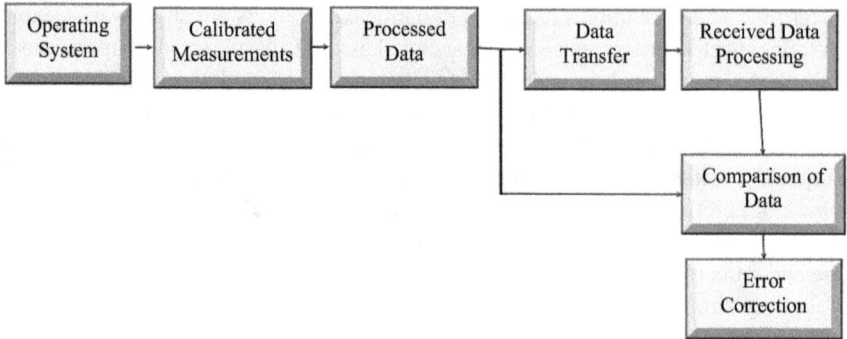

FIGURE 2.5 Block diagram of an end-to-end measurement system. End-to-end measurements involve sensors, processing units, data collection, data transmission, and processing of the data at the receiving end. Therefore, end-to-end calibration requires a ubiquitous outlook on the operation of the systems.

2.11 HARDWARE-BASED CALIBRATION

Hardware-based calibrations have been practiced for many decades, and it is a well-established discipline. Calibration techniques are widely varied, and some techniques are standardized. Some methods will briefly be discussed in this book.

Decades back, the calibrations and adjustments of measurement devices were conducted in the analog domain by manually adjusting hardware components of devices. Manually adjustable components were resistors, capacitors, amplifier gains, frequencies, switches, filters, trimmers, functionality of operational amplifiers, logic circuits, variations in signal sampling, zero-cross settings, and so on. At the fundamental level, many physical calibrations are still conducted on the analog parts by adjusting hardware components not manually but by software commands. The software commands may be generated by dedicated algorithms or AI. For example, in some applications, calibration is realized by software to change the operational characteristics of devices by programmable hardware components, as in the case of loop gain adjustments. These changes result in the reconfiguration of bandwidths and introduce low-noise operational characteristics.[B17]

In many cases, front-end electronics contain embedded controllers, programmable microprocessors, digital signal processors, field programmable gate arrays, data acquisition (DAQ) systems, digital-to-analog converters (DACs), and analog-to-digital converters (ADCs). These front-end electronics are often used for calibration purposes. In some applications, the DAQ of the system is designed around numerous dedicated boards that give control of conversion rates. In other cases, two or more boards are used; while one board manages the system, the others manage the data conversion, interface, calibration, data flow, and debugging. Dedicated boards add further costs, but yield software-based electronic adjustment of components, making calibrations easier.[B18]

AI and programmable hardware-based systems with dedicated on-board calibration facilities allow for calibrations by altering parameters. A typical example is the linear ramp generator, which is programmable by software, for adjusting down-stream operations. Some AI-based applications make use of voltages, frequencies, and currents for adjustments by varying capacitance and resistor values with commands issued by the software. DACs are also used to minimize nonlinearities while the conversion is taking place. Adjustments in R-2R DACs have proven to be effective in improving linearities and reaching calibration aims.[B19]

In hardware-related calibration, the use of analog components is preferred due to their scalability and ability to encode data in a single physical quantity, that is, time, electrical current, charge, or voltage. As an example, in ADC-based calibration, the voltages can continuously be adjusted by software until the relative offset errors are reduced to negligible levels. Calibration algorithms are used not only to fix offsets in ADCs but also to fix static gains and feed-forward compensations by engaging a current-controlled oscillator (CCO).[B20]

Microchips, particularly CMOS-based ICs, attract much attention since they can provide convenient calibration techniques. An RF CMOS-based transceiver is illustrated in Figure 2.6. For example, a chopper-stabilized amplifier operation performs offset correction using switching operations with the internal clock. Among other methods, software-adjustable resistors can be used for reducing offset voltages.[B21]

Some ICs are designed to have benchmarking as well as a tunable circuitry. Tunable and programmable circuit boards are used extensively in AI-based and other algorithm-based calibrations. The benchmarking circuit includes target components and internal reference components. The internal reference component exhibits a lower sensitivity to the changes in test conditions than the target component. The tunable circuitry, which is suitable for calibrations, can include the following: (1) tunable resonant tank, which includes a tunable capacitor network; (2) tunable resistor bank and the resistance of each unit resistor; (3) power amplifiers, which include a tunable current scaling network; (4) bias generators, which include a tunable resistor network; (5) tunable capacitor bank, which enables the calibration of capacitance of the target component; (6) tunable current scaling transistor bank, which varies as conditions change; and (7) current supply control units to alter the currents as needed. In most cases, on-chip resistance and capacitance can be used for calibrating without the external reference resistors.[B23]

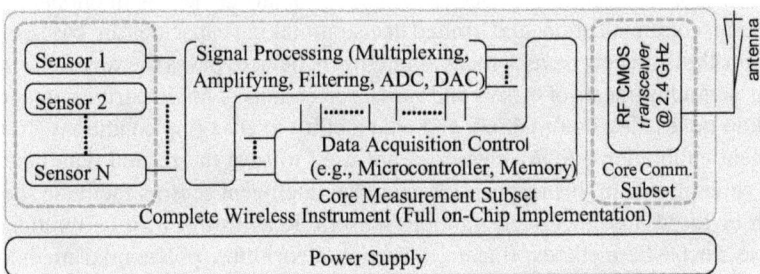

FIGURE 2.6 Schematic illustrates the integration concept of the radio frequency CMOS transceiver, transducers, and electronics in the same microsystem. This device operates at 2.4 GHz and is designed for standalone wireless instruments in biomedical applications. (Source: I. P. Carmo and J. H. Correria.[B22])

In some applications, enhancing high-bandwidth memory access is important. For this, resistor-tuned offset calibration and in situ margin detection techniques are employed. Supply noise adaptation algorithms, coupled with high-accuracy digital delay sensors, enhance voltage stability and mitigate the degradation in supply voltage variations.[B24]

AI-based systems offer additional advantages in calibrations and compensation since they are trained to predict various extra parameters such as power levels, frequency variations, and so on. Compensation can be done in several ways such as nonlinear compensation that linearizes the relationship between input and output, cross-sensitivity compensation due to ambient conditions, and time-based or long-term compensation due to degradation of sensors or their elements.[B19]

Considerable research and design efforts are concentrated on the application of quantum computing in calibrations to enhance remote operations that offer novel techniques, algorithms, and related hardware solutions. However, a significant obstacle is scaling up quantum computers to improve heat properties between cryogenic qubits. A few millikelvin in temperature variations can significantly affect operations.[B25]

Neural networks provide an advantage in computational prediction speed compared with other traditional methods. In many cases, NN can be integrated into common processors on mobile phones, tablets, and PCs, thus paving the way for cost-effective distributed calibration implementations. Also, data-driven and machine learning approaches aim to overcome some of the shortcomings of analytical methods in the calibration of devices with a wide range of variables operating in complex applications.[B26]

2.12   INTELLIGENT CALIBRATION AND SELF-CALIBRATION

Measurement technology has evolved from initial single-device calibrations to integrated, intelligent, and networked calibrations. Intelligent calibrations are conducted on programmable devices such as sensors, instruments, communication networks, and virtual sensors. Intelligent devices have evolved from the concepts of relatively simple smart devices by the addition of onboard AI modules capable of self-calibration and operational decision-making. Smart devices with onboard signal processing, communication, and limited decision-making capacities are based on various algorithms. Recent development of intelligent devices paves the way for self-calibration, self-adjustments of offsets and gains, corrections of nonlinearities, integration of lookup tables, decision making, and adaptability to changing conditions. In many intelligent calibrations, ROM memories are used to save the related data tables for future calibrations. In this respect, for example, intelligent sensors facilitate the calibration of fused and integrated multiple sensors. Readjustment for calibrations can also use stochastic methods, linear calibration algorithms, polynomial methods, as well as AI.

The development of intelligent devices involves the design of reconfigurable systems capable of working with different and complex input signals. Errors, offset, nonlinearity, hysteresis, and cross-sensitivity are rectified by using microprocessors or suitable electronic devices. On board, intelligent algorithms can select multiple

Calibration Algorithms and AI Commands ←

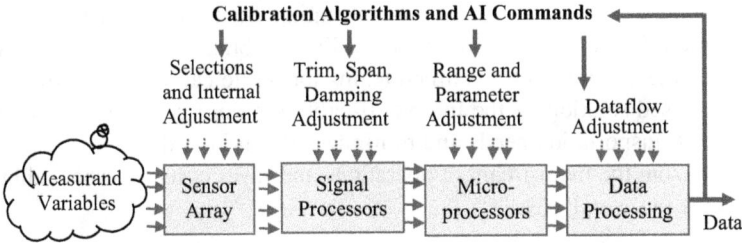

FIGURE 2.7 Block diagram of calibration of a sensor array. Signals from the sensors are processed using adjustable electronic components such as amplifiers, filters, and signal converters. Microprocessors control the internal operations of the measuring device. Calibration algorithms and AI are used to adjust various building blocks of the device.

calibration points from the input-output data. Informed selection of calibration points makes the calibration process simple, accurate, less time-consuming, and low on computational loads.[B19] A typical example of an intelligent sensor array is shown in Figure 2.7.

Intelligent devices conduct calibration by scaling the incoming data, computing statistical information, and communicating with other digital systems on the networks. Some commercially available software permits uploading the new parameters directly to the circuitry of devices. Devices complying with IEEE 1451.4 standards provide comprehensive Transducer Electronic Datasheets (TEDs) that contain information on configuration, scaling, and calibrations.[B27]

In some intelligent sensor applications, each sensor has a memory that is programmed at the factory with a set of default zero and span curves defining the sensor's relationship with the physical phenomenon. These default curves represent average sensor output adjusted for the most accurate response. Each time calibration is done, the appropriate zero or span curve is adjusted in the vicinity of the calibrated values.

Many algorithms have been developed for self-calibration to address a specific family of devices. These algorithms are based on either lookup tables or mathematical models such as linear, second-order, or higher-order nonlinear models. Some applications have multivariate or compounding variables. When solving calibration requirements becomes difficult using statistical or deterministic formulae, self-calibration algorithms based on AI are used. AI models are applied in a variety of tasks like regression, classification, clustering, interpretation, diagnosis, modeling, and control.[B28]

In addition to calibrations, self-calibration algorithms can fix various specific problems in measurements. An example is Levenberg–Marquardt Back Propagation ANN (LMBP-ANN) used in self-calibration. LMBP-ANN collects data on a real-time basis. Once trained, it can compensate for hysteresis, variation in gains, and various nonlinearities while the system is fully operational.[B3]

2.13   INTERNET CALIBRATIONS

Regular calibration is essential to maintain the quality and traceability of devices to national and international standards. One way of achieving traceability is to send instruments to calibration service providers or to qualified accredited organizations.

But this exercise can be costly and time-consuming. Some devices may be bulky for transportation, hence requiring online or offline calibrations on-site. Alternatively, outsourcing the calibration to qualified teams for onsite calibrations. However, one of the most used solutions is the Internet (i-calibration or i-cal). Internet calibration reduces the transportation needs and minimizes downtimes due to the disassembly of a device from the mainstream of operations and saves costs. Internet calibrations can be made in several ways[B29]:

1. Access to the devices can be provided via the Internet using appropriate communication links to exchange information, and
2. Access can be directly provided by telemeters and private networks.

Apart from establishing appropriate communication links, a successful calibration relies on available standards and related software suitable for the Internet calibrations. Some of the basic requirements are given as follows:

1. Computer(s) connected to the internet
2. Relevant calibration software that consists of client programs and service provider's programs
3. Device connection protocols and procedures
4. Availability and implementation of relevant measurement and calibration standards
5. Understanding the principles and functionality of the instrument in calibration
6. Expert personnel at both ends for implementing calibration steps and visual observations
7. Implementing the calibration procedure and repeating the process as many times as needed
8. Conducting mathematical analysis to determine accuracy and uncertainty
9. Generation of related documentation

The internet calibration technique is supported by web-accessible test procedures and appropriate hardware and software. The device under calibration is linked securely via computers to the service provider. Certain steps need to be observed and implemented for successful calibration (Figure 2.8). The iCal service provider instructs the client (i.e., in-house operator) to perform the measurements within a procedural sequence. Once the process is completed, the system generates the required data ascertained from the measurand. Depending on the agreement of the parties involved, the data on calibration can be uploaded to a database. A calibration certificate is generated automatically.

In the implementation of iCal, compliance with the guidelines of any new international standards or any changes in the existing standards may be guaranteed. If requested, the new data and historical records are automatically kept in the clouds and can be accessed and checked by authorized users.

An example of iCal is the calibration of a microwave transmission line. Once the link is established, the client enters the required measurement parameters and is offered options by the provider based on the knowledge of the client's equipment.

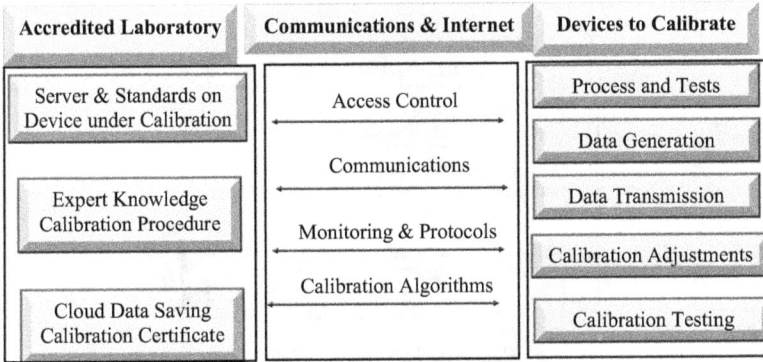

FIGURE 2.8 Block diagram of an internet-based calibration. Internet-based calibrations gained momentum for convenience and cost savings. Devices can be calibrated remotely while operating in industrial locations or in inaccessible areas with the aid of wireless communication networks.

A sweep frequency is generated for measurements of the transmission and reflection coefficients. Calibration is performed using instrument firmware and a set of standard devices, all of which are assumed to be ideal and are available as standard items from the manufacturer. The correction of the measurement data comes via precision verifications, air-spaced transmission lines, attenuators, and terminations, and all the related variable properties can change over time. Once the calibration is completed, all correction factors are stored in an online database.[B30]

Examples of internet calibration are provided by Fluke, UK. This company offers calibration systems for the Fluke 5790B multifunctional instruments.[B31] Anritsu is active in using portable OTDR MT9085 for calibrations.[B32] The OTDR is controlled via the internet with the aid of appropriate software. The software can be controlled through a PC via modem, mobile telephone access, or PCI cards. Similar practices are applied by the National Institute of Standards and Technology (NIST) to realize internet-based calibrations, particularly aiming for radioactive related processing industries.

In some cases, calibration algorithms and production automation know-how are available online, combining standard low-cost integrated circuits and internet access. Online calibration is supported by appropriate software that includes the mathematical models of the interface electronics and optimizes calibration based on knowledge of that model.

One of the first iCal services was established in the UK, National Physical Laboratory (NPL). It aimed for a combination of technologies of remote monitoring, remote control, and calibration standard and techniques. Since then, there have been many Internet calibration providers, particularly in the process industry, health and medical applications, and communications networks. Recent progress in communication systems and the convenient use of the internet have expanded iCal services all around the world.[B33]

During internet or remote calibrations, new problems may arise, such as loss of data, cybersecurity, and compromise with internet safety. Therefore, calibrations need to be conducted safely, and measurement results must be protected reliably.[B34]

FIGURE 2.9 Block diagram of laboratory-based calibration. Many calibration systems still rely on laboratory calibrations for traceability, accreditation, legal requirements, and safety reasons. Laboratories aim for the first degree of calibration traced to the SI standards. Some laboratory-calibrated devices can be used to calibrate other devices within the working standards.

2.14 LABORATORY-BASED CALIBRATION

A significant proportion of calibration techniques require laboratory calibration, which involves measuring the response of the device under laboratory conditions and implementing the necessary corrections. Laboratory calibration relies on the well-maintained and effective use of the standard reference equipment. This requires continuous maintenance in the updating of the equipment, as in the case of chemical, medical, and biological laboratories. Figure 2.9 shows the benefits of using laboratory calibration.[B35]

Laboratory-based calibration is a consistent method to calibrate measuring devices. However, laboratory conditions may not fully represent a wide range of operational conditions that devices may be subjected to intended use. Changes in conditions can cause deviations between the laboratory data and real-time observed data. Device operations inevitably change from time to time, and the devices need to be returned to the laboratory to be calibrated again. The alternative is to take reference calibration equipment to the devices in the field and implement one-point, two-point, span, and multi-point curve fitting calibration as the need arises.[B36]

2.15 LARGE-SCALE AND MASSIVE-SCALE CALIBRATION

Large-scale calibration is the calibration of a large number of devices (thousands, even millions) sequentially or simultaneously. Large-scale calibration is now possible due to the availability of advanced CPSs, AI models, advanced algorithms, communication networks, programmable devices, better knowledge about processes, and other calibration-related technologies.

Large processes entail a series of subprocesses aggregated together to achieve an overall goal. Some examples of such operations with integrated subprocesses are in manufacturing, space exploration, research and design, environmental monitoring, and intelligent transportation systems. Measurement devices in each subprocess need regular calibrations for the overall healthy operations of the processes. The purpose of calibration is to eliminate time-depending factors and total errors supporting the entire process.[B37]

Sensing technology plays an important role and is widely used in critical areas such as battlefield surveillance, environmental monitoring, traffic control systems, medical and healthcare systems, industrial applications, personal health monitoring, space exploration, and agricultural practices.[B38] Large-scale calibrations pose highly complex and challenging issues in parameter determinations, approximations, and univocal mathematical formulations. This necessitates a unified approach in calibrations to be able to include the effects of devices containing hundreds and thousands of parameters. An example of large-scale calibrations is the IoT. IoT connects thousands of devices via the internet and via a range of communications networks. Measuring devices operate individually in a linked form in unison to achieve a purpose. An example of large-scale arrangement is depicted in Figure 2.10.

Some massive-scale calibrations involve hundreds and thousands of sensors spread all over the world in many applications such as environmental sensing, oceanic studies, IoTs, aviation, autonomous fleets, intelligent transport systems, and so on. Periodic recalibration of sensors within these systems is necessary. Clearly, calibrating hundreds of thousands of sensors individually may not be feasible and can be very costly. In most applications, sensor calibration is realized at the backend by collecting data from similar operational sensors and/or the closest reference stations with known and reliable operations. Opportunistic and collaborative sensor calibration is used to achieve accuracy levels close to regulatory standards. In this respect, AI and various deep learning models are extensively tried.[B39]

Massive-scale calibration requires robust design techniques, advanced algorithms, and well-designed infrastructure to address the demands of strategic deployments and the use of sensor technology. In an opportunistic but collaborative system, some

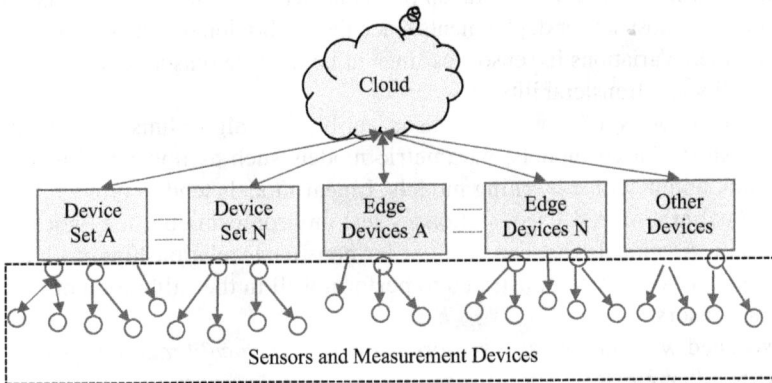

FIGURE 2.10 Block diagram of a large-scale measurement system. Large-scale calibrations can involve thousands of devices. It is computing-intensive, typically using cloud, edge, and fog computing facilities to calibrate geographically dispersed measurement devices. AI is extensively used in large-scale calibrations.

sensors collect calibration information whenever an opportunity presents itself, for example, when the sensor is located close to a reference station. Then the information on calibration is exchanged opportunistically between all others. Recalibration is performed when sufficient information is collected. Information from integrated sensors with different characteristics must be processed appropriately to meet regulatory standards.[B40]

2.16   LOW-COST SENSOR CALIBRATIONS

Low-cost sensors are used in large-scale applications such as environmental monitoring and in some IoT systems. Deployment of a large number of sensors requires different calibration techniques. In large-scale calibrations, some sensors are deployed side-by-side with reference monitors of a certain accuracy. Calibration involves adjusting raw sensor readings using collocation data, suitable mathematical methods, and algorithms.

Issues associated with low-cost sensor performance are inter-sensor and intra-sensor variabilities. Intra-sensor variability occurs due to drift, aging, response time, cross-sensitivity, and sensitivity to environmental factors. The sensor performance is likely to vary for each individual sensor due to drift, aging, response time, cross-response, and environmental factors. *Drift* is the gradual change in sensor response over time, and it may be very different from sensor to sensor. *Aging* refers to the continuous deterioration of sensor performance over time. *Response time* is the time taken (lag) for a sensor to reach a stable reading level. Differences in response times may have serious implications for latency in sampling and data conversion. *Cross-response* is the response of a sensor to other variables rather than the target variable. *Sensitivity to environmental factors* includes the effects of temperature, pressure, humidity, and so forth. Inter-sensor variability occurs due to the differences and variabilities in sensor responses due to intra-sensor variations. Clearly, although multiple identical sensors may be used under the same conditions, possible variations in performances need to be taken into consideration. Inter-sensor variability is crucial for low-cost sensor deployments since the calibration of sensors is dependent on each other. Variations in sensor readings in large-scale sensor deployment affect data viability and transferability.[B41]

Main categories of low-cost sensor calibration algorithms include physical mechanism-based models, parametric models such as linear and non-linear regression, and machine learning models. Linear models tend to provide fast and accurate calibration results in stationary test environments but may not be reliable in situ. AI models that can account for non-linear relationships, such as random forest, have been demonstrated to perform well in the calibration of multiple low-cost sensors.[B42]

Associated with the low-cost sensors is the *reference calibration mapping*. This is a method that creates a reference space from a single sensor and then transforms the output to the remaining sensor population from that reference space. This method results in the ability to utilize low-cost hardware and reduce multiple neural network training needs while meeting the performance requirements.[B43]

Most recent approaches to laboratory control sample calibration are based on AI techniques are mainly linear models such as linear regression (LR) and multi-LR (MLR). There are effective nonlinear models such as RFR, SVR, long short-term memory (LSTM), and CNN. All methods can be evaluated and compared through the mean absolute error, root mean square error, and R^2 scores.[B44]

2.17 MODEL-DRIVEN CALIBRATIONS

Model-based calibration is a classic technique to adjust model parameters using numerical methods to match the experimental data (Figure 2.11). Model-based calibration consists of several steps: (1) modeling and simulation, (2) real-time measurements, (3) diagnosis of original parameters and recalibration probabilities, and (4) compensation. The core step is to establish a model that addresses the requirements of continuity, completeness, and minimization.[B45]

Model-driven methods simplify the real world in descriptive mathematical formats. This approach is highly suitable if the device is not complex and not affected by too many variables. If the device is complex, involving too many measurement variables, it may not be possible to obtain a complete descriptive model. In highly complex situations, data-driven, statistical, and AI methods are employed instead of physical models.[B46]

2.18 MULTIFUNCTION CALIBRATIONS

There are numerous devices that can perform a diverse range of measurement functions. Electric testers are typical examples of such devices, which are capable of measuring resistance, capacitance, inductance, frequency, voltage, and current by the same device. Such devices require calibrations for each function. They are treated like multiple devices within the internationally recognized standards.[B47] Multifunction device calibration finds applications in mechanical, biological, chemical, electrical, temperature, pressure, and flow measurements. Some multifunction devices can measure variables and simulate expected performances. and document the results.[B48]

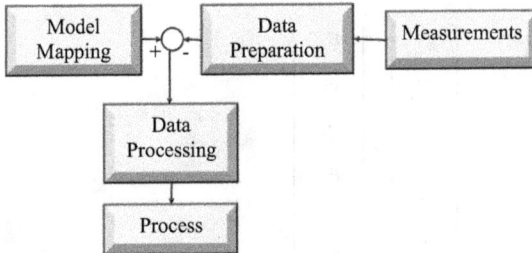

FIGURE 2.11 Block diagram of a model-driven system. In model-driven calibration, ground truth data are prepared and compared with the model-generated data. Discrepancies are noted, and corrective actions are taken. For calibration to be successful, the model of the system and its simulation data must be accurate, descriptive, and comprehensive.

There are many different types of multifunction calibrators offered by a range of companies. Some examples are Fluke 745/754, Omega portable multifunction calibrators, Exetec PRC30, Genii Multifunction Calibrator, and Wavetek 9100 Universal Calibration Systems. Depending on the functionalities, the prices of commercial multifunction calibrators range from US$200 to US$50,000+.

2.19 MULTIVARIATE CALIBRATION

Calibrations can be viewed as univariate calibration and multivariate calibration. In *univariate calibration,* a change in the input variable of a device directly reflects the change in the output response. Therefore, inputs and outputs can be directly correlated, and the inputs and outputs are often related linearly. Most of the univariate systems lead to convenient linear regression analysis. In *multivariate calibrations,* multiple inputs generate multiple independent or interdependent outputs. These outputs may be correlated or uncorrelated with the inputs. Therefore, multivariate calibrations require complex levels of treatment using probabilistic methods, statistical analysis, or AI models. Multivariate measurements and related calibrations find a wide range of applications in biochemistry, chemometrics, and analytical chemistry.[B49]

The matrix description of a multivariate measurement is given in Equations 2.19.1 and 2.19.2. This matrix is a square matrix, but it may not be in many applications.

$$Y = X\beta + \varepsilon \tag{2.19.1}$$

$$
\begin{bmatrix} Y_0 \\ Y_1 \\ Y_2 \\ Y_3 \\ \vdots \\ Y_N \end{bmatrix}
=
\begin{bmatrix}
\beta_{00} & \beta_{01} & \beta_{02} & \beta_{03} & \cdots & \beta_{0N} \\
\beta_{11} & \beta_{12} & \beta_{13} & \beta_{12} & \cdots & \beta_{1N} \\
\beta_{21} & \beta_{22} & \beta_{23} & \beta_{22} & \cdots & \beta_{2N} \\
\beta_{31} & \beta_{32} & \beta_{33} & \beta_{32} & \cdots & \beta_{3N} \\
\vdots & \vdots & \vdots & \vdots & \cdots & \\
\beta_{N1} & \beta_{N2} & \beta_{N3} & \beta_{N2} & \cdots & \beta_{NN}
\end{bmatrix}
\begin{bmatrix} X_0 \\ X_1 \\ X_2 \\ X_3 \\ \vdots \\ X_N \end{bmatrix}
+
\begin{bmatrix} \varepsilon_0 \\ \varepsilon_1 \\ \varepsilon_2 \\ \varepsilon_3 \\ \vdots \\ \varepsilon_N \end{bmatrix}
\tag{2.19.2}
$$

Where $Y_0 \ldots Y_N$ are the outputs, $X_0 \ldots X_N$ are the outputs, $\beta_{00} \ldots \beta_{NN}$ are input parameters, and $\varepsilon_0 \ldots \varepsilon^N$ are the errors.

The first fundamental step toward a successful multivariate calibration is the identification of the relevant explanatory variables that are likely to influence the outputs or introduce errors. In some cases, these variables may be separated and observed individually, whereas in others, all or some of the variables are not separable and they are intertwined. AI models are found to be very effective in calibrating multivariate cases. Application of AI methods such as principal component analysis and distributed regression kernel methods is found to be beneficial in multivariate linear and nonlinear regressions.[B50]

In many circumstances, AI in multivariate calibration methods is easy to implement, but in other methods, they are difficult to implement since they require a robust set of data to be able to isolate the effects of unknown or unaccounted variables. Multivariate calibrations may require substantial resources, research, and time allocation; therefore, the application in measurements is not fully mature yet.[B51]

2.20 ONE-, TWO-, AND MULTIPOINT CALIBRATION

A one-point calibration method is used to correct device errors based on a single point of measurement. An example is if, in a flowrate measurement, the system maintains the same flow continuously, then a single calibration on that flowrate is sufficient. Another example is, if the measured output is linearly related and the slope of the describing curve does not change over the measured range, then it is sufficient to calibrate one point in that measurement range.

A one-point calibration can also be used as "drift checks" to detect changes in response and/or deterioration in system performance. For example, thermocouples used at very high temperatures exhibit a faster "aging" effect. This can be detected by performing periodic one-point calibration and comparing the results with the previous calibrations. One-point calibration leads to *zero span calibration* when the adjustment of one point applies to all the points across the range.[B52]

Two-, three-, and multipoint calibrations apply to situations where different drifts and errors occur at various points throughout the range, and then the number of necessary calibration sections determines the adjustment procedures.[B53]

Multipoint calibration (conventionally four or more points) is used when a device operates over a large but distinguishable range. If the operation is highly nonlinear, the nonlinearity can be broken into smaller linearly expressible sections; each section requires dedicated one-point or multipoint calibration methods (Figure 2.12). Multiple distinct reference points are used for adjustments.

In many cases, calibration requirements can range from a single variable to multiple variables. In these cases, multipoint calibration techniques are employed by identifying and calibrating each variable while keeping the other variables constant.

2.21 PHYSICS-BASED CALIBRATION

Physics-based calibrations use fundamental physical properties and biological or chemical principles. The fundamental laws are used for a mathematical description of the properties of the measurand and the measurement devices (Figure 2.13). Basic physics-based models are proportional models, inversely proportional models, linear models, polynomial models, exponential models, oscillatory and harmonic models, sudden change models, and many others.

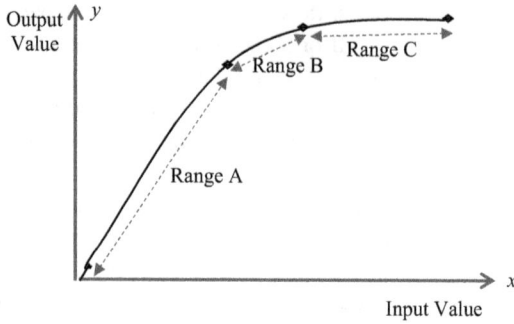

FIGURE 2.12 Multipoint calibration. Multipoint calibration is used if the response of the device to a measurand or the measurand itself is nonlinear. The input–output response of the measuring device is divided into manageable linear sections, and each section is calibrated separately. In a typical multipoint calibration implementation, the controlling elements (i.e., microprocessors) are programmed to identify each section of operations and implement the correct calibration and control procedures on the device.

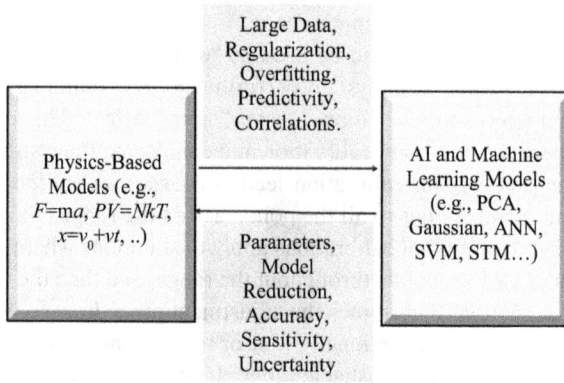

FIGURE 2.13 Physics-based calibration. Physics-based calibrations make use of the mathematical description of the operation. The mathematical descriptions are compared with the AI-generated predictions to assess the differences and implement the correct calibration processes. Here, reliable and trustworthy AI predictions and a good physical description of the system are necessary.

Physical parameters that affect a process are identified first, and then models are created based on these parameters. Some typical examples of physics-based systems are low-cost sensors, various mechanical systems, quantum mechanics, spacetime, gravity, electromagnetic wave propagation, properties of light, atomic physics, and so on. Once the model is identified, simulations can be conducted to describe the process. Simulation-generated data can then be used with AI models. In many applications, creating descriptive models with certain accuracy can be highly complex and difficult, thus needing the use of more advanced techniques.[B54]

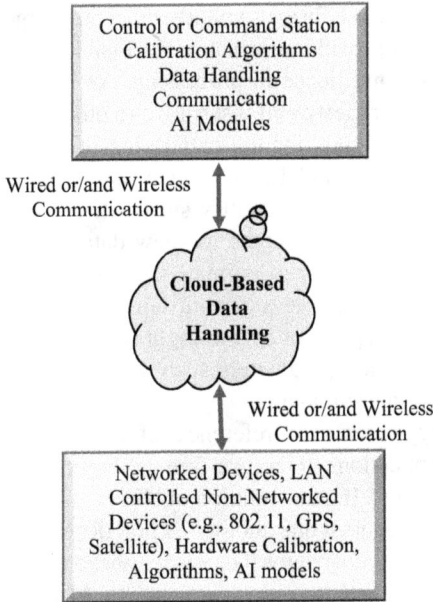

FIGURE 2.14 Remote calibration. Remote calibrations are an essential part of the calibration of devices in many operations, such as space stations, satellites, planetary devices, space telescopes, and so on. Remote calibration technology is highly developed, and various AI models are extensively used.

Physics-based models can be computationally complex. AI models are extensively used in physics-based calibrations, some of the models are SVM, LSTM, brain-inspired HDC, and others.[B55]

2.22   REMOTE, FIELD, AND ON-THE-FLIGHT CALIBRATION

Remote calibration is the process of calibrating a measuring device on-site (termed as on-the-flight calibration). Various forms of communication methods are used, ranging from the internet and direct radio links to microwave and satellite communications (Figure 2.14). Compared to laboratory-based calibrations, remote calibration can eliminate installation errors and minimize systematic errors. Nevertheless, at the same time, it is likely to introduce its own errors, such as data loss, latency errors, maliciously introduced errors, and other errors inherent in communication systems.[B56]

2.23   SENSOR CALIBRATION

The use of sensors is ubiquitous in modern society; consumer electronics, smartphones, wearables, healthcare systems, industry, and autonomous cars are a few examples to mention. However, sensor reading may be corrupted by noise and by deviations from their ideal characteristics.[B57,B58]

A sensor is defined as "a device or subsystem whose purpose is to detect events or changes in its environment and respond with an electrical signal, where the output signals of sensors may be in the form of voltage, current, charge, light, magnetic, and electromagnetic." Sensors provide information accessible to electronics components and digital systems for convenience in processing. An ideal sensor should only be sensitive to the measured property and insensitive to other affecting factors.[B59]

There are numerous methods for the calibration of sensors, such as statistical estimation algorithms and AI models. In a single sensor calibration, algorithms use physical modeling of the sensor to identify similarities between the recorded evidence and the physical operations. Once the new data are obtained from the sensor, the data are processed suitably to estimate the relevant parameters by statistical methods or AI models.[B60] In large and multisensory systems, the state-of-the-art technique of calibration is in place or on-the-flight (also termed as macro-calibration) network-wide. The calibration parameters such as gains, biases, and drifts can be estimated using on-field measurements.

In some multisensory cases, prior reference information may not be available for calibration. In such applications, reference-free calibration methods can be used by blind calibration algorithms. Implementation of blind calibration algorithms leads to the calibration of sensors against one another. A practical limitation of blind calibration is the assumption of homogeneity. The sensors need to be densely deployed and operated in the same homogeneous environment.[B61] AI models are applied extensively in multisensory applications. Some AI models are embedded on board the device and/or implemented by cloud or by edge devices.[B62]

2.24   SIGNAL-BASED CALIBRATIONS

Some devices (such as voltmeters, ammeters, liquid flowmeters, and oscilloscopes) are calibrated by injecting suitable continuous voltage or current signals. The most used calibration signals are square waveform, sinusoidal waveform, and sawtooth waveform, as illustrated in Figure 2.15.

A typical example of calibration devices is the OMEGA™ CL427 indicator/simulator. It is a portable multifunction calibrator and a function generator used to calibrate electromechanical devices, counters, relays, 4–20 mA systems, and thermocouples. It generates sine, square, and triangular waveforms in the range of millivolts, volts, and milliamperes with frequencies 0.3 Hz to 20 kHz. Pulse heights of signals are adjustable between 100 mV and 20 V. Other similar calibrators are Uni-T UT 745 pressure calibrator, Kingsine KS823 standard source meter calibrator, Time Electronics 5051Plus multifunction calibration system, and thousands more other products.[B63] [B64]

2.25   CALIBRATION CONCEPTS IN SOCIAL SCIENCES

Calibration is applicable in situations where changes occur in a dependent variable (output) due to some influencing factors (inputs). The only condition is that they should be observable and measurable. Dependent and observable variables are also encountered in social sciences, economics, and life sciences. So long as these variables are monitorable and measurable, then appropriate calibration methods can be applied to independent variables to achieve desirable outcomes.

Calibration-related topics in social sciences are widely discussed and highly applicable concepts. In social sciences, one of the definitions of calibration is "being

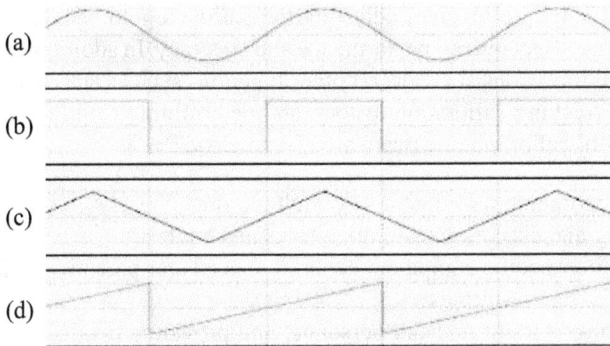

FIGURE 2.15 Signals mostly used in signal-based calibration are (a) sinusoidal waves, (b) square waves, (c) triangular waves, or (d) sawtooth waves generated by external devices. Signal-based calibrations are suitable for many electrical and mechanical measurement devices such as voltmeters, oscilloscopes, counters, ammeters, flowmeters, density gauges, etc. These devices are injected with suitable signals recommended by the manufacturers. Calibration is done manually through adjustable components such as resistors and capacitors.

able to make appropriate responses to social cues or the social environment". Many calibration techniques discussed in this book are also applicable in social science and life science. Although mathematical treatments are the same, the measurable variables and measuring techniques are different. Due to the availability of space, only a few examples will be provided in this book.

In economics and finance, the mathematical calibration models and techniques discussed here are used to assess the validity and success of policy decisions. These policy models (e.g., Bayesian methods) are some of the tools to determine the health of the economic outcomes. Well-developed models can combine multiple sources of evidence about how policy alternatives can impact those outcomes and synthesize outcomes into measures salient for the policy decision. Calibrating these policy models to fit empirical data provides validity and improves the quality of model predictions. These methods lead to policy decisions by balancing distributions of model parameters or altering fundamental structural assumptions or readjusting the likelihood functions obtained from the calibration data.[B65] [B66] [B67]

In marketing, calibration is used for selecting model parameters to ensure that the theoretical value of the model describes and matches the market prices as closely as possible at any given time. In investments, calibration is the process of using observed transactions in a portfolio using own instruments and transactions in which funds enter a position. This ensures the valuation techniques employed to determine the portfolio of investment.[B68]

2.26 SOFTWARE IN CALIBRATION

Software support is an essential part of the calibration of measurement devices. Calibration adjustments can be made by software to ensure the effectiveness and reliability of calibration. Related software can be supplied by third parties, or it can be one-off specialized software solely dedicated to a particular device or a process.

Calibration software involves adjusting the parameters to alter device readings. Therefore, it can affect overall performance and accuracy. In addition to calibrations, many software tools consist of several other functions such as automated calibration management, tracking calibration history, and determining calibration frequencies. The functions of calibration software are given below:[B69] [B70]

- Gathering data in real-time across multiple sites
- Planning and executing calibration tasks and analysis
- Scheduling specific equipment calibration needs for proactive maintenance and reducing downtimes
- Delegating tasks to relevant personnel and providing details on the use of selected standards
- Maintain data history and certificate generation
- Using mathematical tools for analysis and decision making

An example of calibration software is for wireless network applications. A Python package provides an object-oriented library for RF network analysis, circuit building, calibration, and network simulation. This software assists microwave network operations, conducts frequency/port slicing, network analyzer calibrations, time gating, vector fitting, interpolating between set of networks, deriving network statistical properties, plotting of rectangular plots (decibels, magnitude, phase, and group delays), plotting Smith charts, and uncertainty analysis.[B71]

Another example is the platform Deep Underground Neutrino Experiment (DUNE) used in CERN. In this research establishment, the Photon Detection System (PDS) is based on photomultiplier tubes and allows triggering on the scintillation light signals produced by cosmic rays and other charged particles traversing the detector. The acquisition and calibration software controls the high-voltage power supplies, the calibration system, and the PDS DAQ. In the calibration mode, it allows for the selection of acquisition trigger mode, controls and defines the acquisition settings, selects the trigger points, arranges the calibration mode and settings, and provides a graphical user interface. For DAQ, the calibration mode grants access to some of the front-end settings. The program user can choose channel groups to be acquired, sampling periods, window sizes, and post-trigger samples.[B72]

There are numerous calibration software programs tailored for specialized applications. Some examples are "Reftab" for asset management,[B73] "Metquay" for calibration laboratory management,[B74] "QT9QMS" for automation of calibration, managing calibration records, and calibration auditing,[B75] and hundreds of others.

2.27 STATIC AND DYNAMIC CALIBRATIONS

Calibrations of measuring devices can be made under static or dynamic conditions. *Static calibrations* are usually conducted offline, and the variations in the input variables are not time-dependent. *Dynamic calibrations* are conducted while the device is fully or partially operational, and the variations in the input variables can change in time. The performance characteristics of the device in dynamic conditions may be significantly different from the performance under static conditions. In dynamic calibration, many other influencing factors on the input variables may occur. Some of these influencing factors are pressure, temperature, vibrations, radiation, drifts, biases, and unaccounted interferences.[B76]

The term "*static*" refers to a calibration procedure in which the values of the influencing factors remain constant (or can be made constant), that is, they do not change in time. Measured calibration points describe the static input–output relationship for the system. Often, linear or polynomial curves describe the relationship as

$$y = f(x) \qquad\qquad (2.27.1)$$

Where x is the input variable, and y is the output variable.

Static calibration dictates that a change in input quantity causes a corresponding predictable change in the output close to a calibration standard. As a general practice, the calibration standard should demonstrate several times better accuracy than the measured values of the device.

Static performance characteristics include linearity, static sensitivity, repeatability, hysteresis, resolution, and the readability of the results. For example, in mechanical systems, hysteresis plays an important role in calibrations. Hysteresis may be caused by internal friction, sliding friction and heat, external friction, and free play mechanisms. Hysteresis can mathematically be accounted for by taking readings in ascending and descending order of the inputs and by calculating arithmetic means of the output values.[B77]

In real-world calibrations, linearity cannot fully be achieved. The deviations from the ideal values are referred to as linearity tolerances. For example, 2% independent linearity means that the output will remain within ±2% of the full-scale output from the standard (idealized) line. In many applications, nonlinear responses can still be approximated to a linear form over a restricted range. Static sensitivity in linear systems can be viewed as the slope of a calibration curve.

Some multiple input/output calibrations may allow for observation of each input and the corresponding output in isolation. Then, individual inputs are varied in increments in increasing and decreasing directions over a specified range. The observed output then becomes a function of that single input. This procedure is repeated for all other inputs one by one, thus developing a family of relationships between the inputs and outputs. Clearly, each input will yield its own related output values.[B78]

In multivariate static calibrations, the input/output relationship usually demonstrates statistical characteristics. From these characteristics, appropriate calibration curves can be obtained, and various statistical techniques can be applied.

Dynamic calibration variables may be time-dependent in their magnitudes and frequencies depending on the situations in which the devices are operating. The causal relationship between the input and output of a measuring device may be expressed as.[B79]

$$y = \left(x_1, \ x_2, x_3, \ x_4, \dots x_n \ \right) \qquad\qquad (2.27.2)$$

When AI is used, the terms dynamic and static calibrations take slightly different meanings. In static calibrations, data are collected while the system is operating in isolation from the intended application. Dynamic calibration indicates that the device under calibration is integrated fully into the system and operates with the other instruments. The data collection and data preparation in static and dynamic conditions may be quite different.

In an application, the parameters of ordinary and partial differential equations are determined by an off-the-shelf ANN model. The model output was then compared with the ground truth data and the data obtained from numerical simulations. A loss function is defined to quantify the mismatch between the ANN prediction and the ground truth over a predefined window. This loss function is then minimized using methods like gradient descent optimization using a back-propagation algorithm. This *dynamic* approach in the training of the model is contrasted with a *static* calibration result by mathematical testing algorithms such as the least square regression estimates.[B80]

In an AI application, static and dynamic calibration aims to determine if the measuring device is stationary or it is operating in a dynamic environment. For example, accelerometers are used to measure the motion of certain dynamic objects to assess the intensity of the operational state of the system.[B81] Another AI application attempts to design a method for an automatic calibration system based on dynamic load testing subroutines and dynamic scheduling hardware. Dynamic loading technology is dependent on a uniform interface, which can make the system more open and easier to maintain and update.[B82]

2.28   TARGET CALIBRATION

The term *target calibration* is used in a few different ways. For example, target calibration is interpreted as several different types of measuring devices that are used to achieve a single outcome in a process. Each device is separately calibrated to achieve the best overall results. However, they also need to be calibrated in unison to address the target measurement (collated measurement) of the system. Calibrating each device requires its own standard or reference. Their collated results introduce new requirements to obtain an optimal calibration of the target. In one application, elemental composition of samples, molecular vibrational features, fluorescence, morphology, and the texture provide a full picture of the sample with complete information that needs to be co-aligned, correlated, and individually calibrated.[B83]

An example target calibration is SuperCam, which is an integrated remote-sensing instrumental suite for NASA's Mars 2020 mission. It consists of a co-aligned combination of laser-induced breakdown spectroscopy, time-resolved Raman and luminescence, visible and infrared spectroscopy, together with sound recording (MIC) and high-magnification imaging techniques. Together, they provide information on geochemistry and mineral contents around the Perseverance Rover.[B84] [B85]

Another example is the measurement networks for greenhouse gases, which contain multiple variables to obtain a target result. This network is vital for understanding global emission trends and their effectiveness in determining emission mitigation

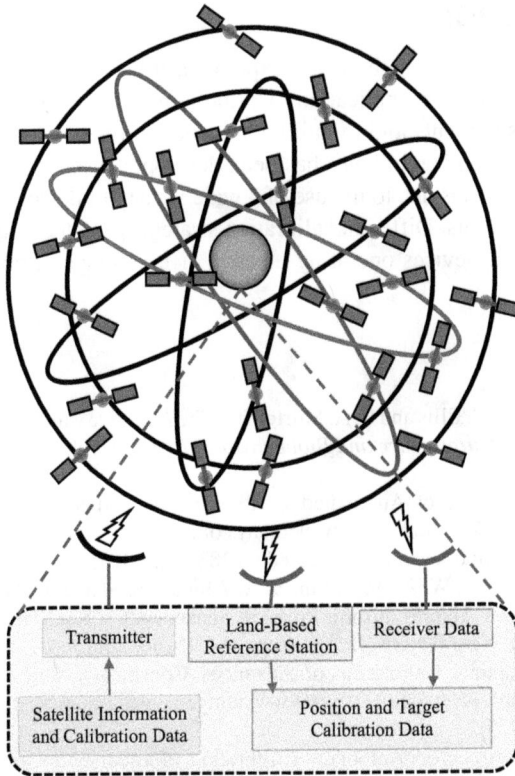

FIGURE 2.16 Satellite-based calibration of user devices. Satellites are used in many applications from location determination to pollution monitoring. They are calibrated by suitable remote calibration mechanisms. In location determination, calibration of devices of the end-users is conducted using signals picked up from the satellites and from the land-based reference stations.

policies, strategies, and initiatives, and ascertaining the effectiveness of emission reduction targets at the local, regional, and global scales.[B86]

Another interpretation of target calibration is the determination of the position of an object, shown in Figure 2.16. In this geostationary satellite system, each satellite is calibrated for correct position, information transmission, and timing. Geostationary satellites are essential for communications, navigation, satellite TVs, data relays, disaster and emergency response, GPS, military applications, and weather monitoring. Geostationary satellites play an important role in global positioning systems. They make precise location determination, location tracking, and navigation. Signals picked up from satellites are calibrated on user devices. Precise position calibration employs methods like two-point or multi-point reference stations.[B87] [B88] [B89] [B90]

2.29 CONCLUSIONS

In this chapter, calibration methods are explained. It has been shown that there is a rich repertoire of calibration applications. Calibration technology is used for the traceability of the measurements and to build confidence in the results. A rich volume of calibration strategies and techniques has been developed after having been practiced for many decades. This is due to the use of a wide variety of devices and their diverse operational requirements. Different calibration strategies and techniques that are suitable for the needs of devices operating under a diverse range of conditions must be used.

REFERENCES

B1. M. D. Giudice, B. J. Ellis and E. A. Shirtcliff, "The Adaptive Calibration Model of Stress Responsivity," *Neuroscience and Biobehavioral Reviews*, vol. 35, no. 7, pp. 1562–1592, June 2011.

B2. E. Ring, The Future of Automated Calibration: AI and Machine Learning, Posted on Nov. 2023, https://www.metquay.com/post/the-future-of-automated-calibration-ai-and-machine-learning (Accessed on 18 April 2025).

B3. A. M. M. Almassri, W. Z. W. Hasan, S. A. Ahmad, S. Shafie, C. Wada and K. Horio, "Self-Calibration Algorithm for a Pressure Sensor with a Real-Time Approach Based on an Artificial Neural Network," *Sensors*, vol. 18, no. 2561, pp. 1–16, 2018.

B4. Numgeo, Automatic Calibration of Advanced Constitutive Soil Models (Automatic Calibration Tool – Numgeo). https://www.numgeo.de/automatic-calibration/ (Accessed on 18 April 2025).

B5. Z. Taylor and J. Nieto, "Automatic Calibration of Lidar and Camera Images Using Normalized Mutual Information," www-personal.acfr.usyd.edu.au/jnieto/Publications_files/TaylorICRA2013.pdf (Accessed on 18 April 2025).

B6. T. Barrett and A. K. Mishra, "Statistical Study of Sensor Data and Investigation of ML-Based Calibration Algorithms for Inexpensive Sensor Modules: Experiments from Cape Point," *IEEE Transactions on Instrumentation and Measurement*, vol. 73, pp. 1–10, 2024, Article no. 1003310, 2024.

B7. D. Stucchi, L. Magri, D. Carrera and G. Boracchi, "Multimodal Batch-Wise Change Detection," *IEEE Transactions on Neural Networks and Learning Systems*, vol. 34, no. 10, pp. 6783–6797, Oct. 2023.

B8. J. Zou and O. Petrosian. "Explainable AI: Using Shapley Value to Explain Complex Anomaly Detection ML-Based Systems" 2020. https://www.researchgate.net/publication/344238929_Explainable_AI_Using_Shapley_Value_to_Explain_Complex_Anomaly_Detection_ML-based_Systems (Accessed on 18 April 2025).

B9. X. Fang and I. Bate, "An Improved Sensor Calibration with Anomaly Detection and Removal," *Sensors and Actuators B: Chemical*, vol. 307, p. 127428, 15 March 2020.

B10. R. Tan, G. Xing, X. Liu, J. Yao and Z. Yuan, "Adaptive Calibration for Fusion-Based Cyber-Physical Systems," *ACM Transactions on Embedded Computing Systems (TECS)*, vol. 11, no. 4, pp. 1–25, Jan. 2013.

B11. C. Zhao, Y. Shi, Y. Du, S. Jiang, Y. Ji and X. Zhao, "A Rapid and Convenient Spatiotemporal Calibration Method of Roadside Sensors Using Floating Connected and Automated Vehicle Data," *IEEE Transactions on Intelligent Transportation Systems*, vol. 25, no. 9, pp. 10953–10966, Sept. 2024.

B12. A. C. Marceddu et al., "Air-to-Ground Transmission and Near Real-Time Visualization of FBG Sensor Data via Cloud Database," *IEEE Sensors Journal*, vol. 23, no. 2, pp. 1613–1622, 15 Jan. 2023.

B13. A. F. Villaverde, F. Froehlich and J. R. Banga, "A Protocol for Dynamic Model Calibration," *Briefing in Bioinformatics*, vol. 23, no. 9, pp. 1--19, Oct. 2021.

B14. W. Liang, R. Chen and Y. Zhang "A Dynamic Calibration Method of Free-Field Pressure Sensor Based on Hopkinson Bar," *AIP Advances*, vol. 10, p. 075223, July 2020.

B15. X. Liu et al., "Fusing Physics to Fiber Nonlinearity Model for Optical Networks Based on Physics-Guided Neural Networks," *Journal of Lightwave Technology*, vol. 40, no. 17, pp. 5793–5802, Sept. 2022.

B16. A. Rachman, J. Seiler and A. Kaup, "End-to-End Lidar-Camera Self-Calibration for Autonomous Vehicles," *2023 IEEE Intelligent Vehicles Symposium (IV)*, Anchorage, AK, USA, pp. 1–6, 2023.

B17. I. Akita, T. Kawano, H. Aoyama, S. Tatematsu and M. Hioki, "An Automatic Loop Gain Enhancement Technique in Magnetoimpedance-Based Magnetometer," *IEEE Journal of Solid-State Circuits*, vol. 57, no. 12, pp. 3704–3715, Dec. 2022.

B18. S. Mastroianni et al., "Design and Performance of Data Acquisition and Control System for the Muon g-2 Laser Calibration," *IEEE Transactions on Nuclear Science*, vol. 67, no. 5, pp. 832–839, May 2020.

B19. J. Rivera, G. Herrera, M. Chacón, P. Acosta and M. Carrillo, "Improved Progressive Polynomial Algorithm for Self-Adjustment and Optimal Response in Intelligent Sensor," *Sensors*, vol. 8, pp. 7410–7427, 2008.

B20. R. Khaddam-Aljameh et al., "HERMES-Core—A 1.59-TOPS/mm2 PCM on 14-nm CMOS In-Memory Compute Core Using 300-ps/LSB Linearized CCO-Based ADCs," *IEEE Journal of Solid-State Circuits*, vol. 57, no. 4, pp. 1027–1038, April 2022.

B21. H. Morimoto, H. Goto, H. Fujiwara and K. Nakamura "CMOS Op-amp Offset Calibration Technique Using a Closed Loop Offset Amplifier and Compact Resistor String DAC," *2011 International Conference on Solid State Devices and Materials*, Nagoya, pp. 182–183, 2011.

B22. I. P. Carmo and J. H. Correria, "Wireless instrumentation," In: Editors J. G. Webster and H. Eren, *Measurements Instrumentation and Sensors Handbook: Spatial, Mechanical, Thermal, and Radiation Measurements* (2nd ed.). CRC Press, Boca Raton, USA, Ch. 85 pp, 1–14, 2014.

B23. C. W. Lee, "On-Chip Benchmarking and Calibration without External References," Technical Report No. UCB/EECS-2012–7. https://www.eecs.berkeley.edu/Pubs/TechRpts/2012/EECS-2012-7.html (Accessed on 18 April 2025).

B24. K. Chae et al., "A 4-nm 1.15 TB/s HBM3 Interface with Resistor-Tuned Offset Calibration and In Situ Margin Detection," *IEEE Journal of Solid-State Circuits*, vol. 59, no. 1, pp. 231–242, Jan. 2024.

B25. P. R. Genssler et al., "Cryogenic Embedded System to Support Quantum Computing: From 5-nm FinFET to Full Processor," *IEEE Transactions on Quantum Engineering*, vol. 4, pp. 1–11, 2023.

B26. Z. Ballard, C. Brown, A. M. Madni and A. Ozcan, "Machine Learning and Computation-Enabled Intelligent Sensor Design," *Nature Machine Intelligence*, vol. 3, pp. 556–565, July 2021.

B27. "IEEE Standard for a Smart Transducer Interface for Sensors and Actuators–Common Functions, Communication Protocols, and Transducer Electronic Data Sheet (TEDS) Formats," IEEE Std 1451.0–2024 (Revision of 1451.0–2007), pp. 1–429, doi: 10.1109/IEEESTD.2024.10571828. 26 June 2024.

B28. A. Sinha and D. Das, "sCalib: A Warehouse Sensor Fault Detection and Self-Calibration Technique for Sustainable IoT," *2021 IEEE 18th India Council International Conference (INDICON)*, Guwahati, India, pp. 1–6, 2021.

B29. I. A. Ershov and O. V. Stukach, "Internet of Measurement Development Based on NI PXI Remote Calibration," *2019 Dynamics of Systems, Mechanisms and Machines (Dynamics)*, Omsk, Russia, pp. 1–5, 2019.

B30. S. Madhuwantha, P. Ramabadran, R. Farrell and J. Dooley, "Novel Calibration Technique for Wideband Transmitters Using Constellation Mapping," *2019 92nd ARFTG Microwave Measurement Conference (ARFTG)*, Orlando, FL, USA, pp. 1–4, 2019.

B31. https://www.fluke.com/en-us/product/calibration-tools/electrical-calibration/electrical-standards/5790b (Accessed on 19 April 2025).

B32. https://www.anritsu.com/en-gb/test-measurement/products/mt9085series (Accessed on 19 April 2025).

B33. R. A. Dudley, A. G. Morgan and N. M. Ridler, 2001 "Advances in NPL's Internet Calibration and Measurement Services for High-Frequency Electrical Quantities," *BEMC 2001- 10th British Electromagnetic Measurement Conference,* Harrogate, UK, 6–8 November 2001.

B34. European Association of National Metrology Institutes *Guidelines on the Calibration of Digital Multimeters* EURAMET cg- 15, Version 3, 02/2015. https://www.euramet.org/Media/docs/Publications/calguides/EURAMET_cg-15__v_2.0_Guidelines_Calibration_Digital_Multimeters.pdf (Accessed on 24 July 2025)

B35. F. Cordara, D. Orgiazzi and V. Pettiti, "Interlaboratory Comparisons for Frequency Calibration: A First Two-Year Campaign in Italy," *2013 Joint European Frequency and Time Forum & International Frequency Control Symposium (EFTF/IFC)*, Prague, Czech Republic, pp. 306–309, 2013.

B36. X. Shang, N. Ridler, W. Sun, P. Cooper and A. Wilson, "Preliminary Study on WM-380 Waveguide TRL Calibration Line Standards at the UK's National Physical Laboratory," *2019 92nd ARFTG Microwave Measurement Conference (ARFTG),* Orlando, FL, USA, pp. 1–4, 2019.

B37. D. Xing, D. Xu, F. Liu, H. Li and Z. Zhang, "Precision Assembly among Multiple Thin Objects with Various Fit Types," *IEEE/ASME Transactions on Mechatronics*, vol. 21, no. 1, pp. 364–378, Feb. 2016.

B38. S. Agriesti, V. Kuzmanovski, J. Hollmén, C. Roncoli and B. -H. Nahmias-Biran, "A Bayesian Optimization Approach for Calibrating Large-Scale Activity-Based Transport Models," *IEEE Open Journal of Intelligent Transportation Systems*, vol. 4, pp. 740–754, 2023.

B39. B. Khaleghi et al., "Opportunistic Calibration of Smartphone Orientation in a Vehicle," *2015 IEEE 16th International Symposium on a World of Wireless, Mobile and Multimedia Networks (WoWMoM)*, Boston, MA, USA, pp. 1–6, 2015.

B40. N. H. Motlagh et al., "Toward Massive Scale Air Quality Monitoring," *IEEE Communications Magazine*, vol. 58, no. 2, pp. 54–59, February 2020.

B41. C. Yang, S. Chatterjee and T. J. Oechtering, "Enhancing Network Calibration for Low-Cost Gas Sensor Networks through Adaptive Similarity Search," *ICASSP 2025 – 2025 IEEE International Conference on Acoustics, Speech and Signal Processing (ICASSP)*, Hyderabad, India, pp. 1–5, 2025.

B42. A. Wang, "Leveraging Machine Learning Algorithms to Advance Low-Cost Air 2 Sensor Calibration in Stationary and Mobile Settings" *Atmospheric Environment*, vol. 301, pp. 119692–119692, 2023.

B43. N. Murrell, R. Bradley, N. Bajaj, J. Whitney and G. T. C. Chiu, "New Calibration Method for Implementing Machine Learning in Low-Cost Sensor Applications," *IEEE Sensors Letters*, vol. 4, no. 2, pp. 1–4, Feb. 2020.

B44. S. D. Vito, G. D. Elia, S. Ferlito, E. Esposito, G. Piantadosi and G. D. Francia, "Remote Calibration Strategies for Low-Cost Air Quality Multisensors: A Performance Comparison," *2024 IEEE International Symposium on Olfaction and Electronic Nose (ISOEN)*, Grapevine, TX, USA, pp. 1–4, 2024.

B45. Z. Yan, Y. Wang, H. Liu, J. Xiao and T. Huang, "An Improved Data-Driven Calibration Method with High Efficiency for a 6-DOF Hybrid Robot," *Machines*, vol. 11, p. 31, 2023.

B46. T. Wissel, B. Wagner, P. Stüber, A. Schweikard and F. Ernst, "Data-Driven Learning for Calibrating Galvanometric Laser Scanners," *IEEE Sensors Journal*, vol. 15, no. 10, pp. 5709–5717, Oct. 2015.

B47. K. Tingting, L. Yan, Z. Lei and H. Yan, "Development of Multi-Channel Automatic Digital Multimeter Calibration Device," *2019 14th IEEE International Conference on Electronic Measurement & Instruments (ICEMI)*, Changsha, China, pp. 108–113, 2019.

B48. J. Woo, H. Choi and S. Sim, "Mobile-Assisted Calibration for Multi-Display Devices: Enhancing Color and Brightness Uniformity," *2025 IEEE International Conference on Consumer Electronics (ICCE)*, Las Vegas, NV, USA, pp. 1–4, 2025.

B49. S. Brown, R. Tauler and B. Walczak, *Comprehensive Chemometrics: Chemical and Biochemical Data Analysis* (2nd ed.). Elsevier, London, UK, p. 2944, May 2020.

B50. L. S. Rodrigues et al., "Practical Multivariate Data Correction and Sensor Calibration Methods," *IEEE Sensors Journal*, vol. 20, no. 13, pp. 7283–7291, July 2020.

B51. M. Chvosteková, "A Comparison of Multiple-Use Confidence Regions for Multivariate Calibration," *2019 12th International Conference on Measurement*, Smolenice, Slovakia, pp. 244–247, 2019.

B52. J. Peng, J. R. Piepmeier, S. Misra, P. Mohammed and A. Bringer, "One-Point Calibration for Soil Moisture Active/Passive (SMAP) L-Band Microwave Radiometer," *IEEE Journal of Selected Topics in Applied Earth Observations and Remote Sensing*, vol. 18, pp. 655–662, 2025.

B53. T. Liu, B. Li and L. Yang, "Phase Center Offset Calibration and Multipoint Time Latency Determination for UWB Location," *IEEE Internet of Things Journal*, vol. 9, no. 18, pp. 17536–17550, Sept. 2022.

B54. A. Biswas and M. Cotorogea, "Advanced Physics-Based Compact Models for New IGBT Technologies," *2020 International Exhibition and Conference for Power Electronics, Intelligent Motion, Renewable Energy and Energy Management*, Germany, pp. 1–7, 2020.

B55. P. R. Genssler, H. E. Barkam, K. Pandaram, M. Imani and H. Amrouch, "Modeling and Predicting Transistor Aging under Workload Dependency Using Machine Learning," *IEEE Transactions on Circuits and Systems I: Regular Papers*, vol. 70, no. 9, pp. 3699–3711, Sept. 2023.

B56. J. Song, D. -J. Kim, S. Lee, S. An, J. -H. Hwang and J. Kim, "Geometric Positioning Error Mitigation of SAR Image in Ocean Utilizing AIS Information," *IEEE Geoscience and Remote Sensing Letters*, vol. 20, pp. 1–5, 2023.

B57. Y. Pi, M. Wang, B. Yang and Z. Gao, "Robust Camera Distortion Calibration via Unified RPC Model for Optical Remote Sensing Satellites," *IEEE Transactions on Geoscience and Remote Sensing*, vol. 60, pp. 1–15, Article no. 5627815, 2022.

B58. A. Mikov, S. Reginya and A. Moschevikin, "In-Situ Gyroscope Calibration Based on Accelerometer Data," *2020 27th Saint Petersburg International Conference on Integrated Navigation Systems (ICINS)*, St. Petersburg, Russia, pp. 1–5, 2020.

B59. S. Schuster, J. Wetzel, S. Zeitvogel and A. Laubenheimer, "Automatic Extrinsic Multi-Sensor Network Calibration Based on Time Series Matching," *2024 27th International Conference on Information Fusion (FUSION)*, Venice, Italy, pp. 1–8, 2024.

B60. L. Feng et al., "An Intuitively Derived Decoupling and Calibration Model to the Multiaxis Force Sensor Using Polynomials Basis," *IEEE Sensors Journal*, vol. 24, no. 7, pp. 9514–9522, April 2024.

B61. R. T. Rajan, R. -V. Schaijk, A. Das, J. Romme and F. Pasveer, "Reference-Free Calibration in Sensor Networks," *IEEE Sensors Letters*, vol. 2, no. 3, pp. 1–4, Sept. 2018.

B62. S. De Vito et al., "A Global Multiunit Calibration as a Method for Large-Scale IoT Particulate Matter Monitoring Systems Deployments," *IEEE Transactions on Instrumentation and Measurement*, vol. 73, pp. 1–16, Article no. 2501916, 2024.

B63. Portable Multifunction Calibrator. https://in.omega.com/pptst/CL427.html (Accessed on 18 April 2025).

B64. X. Li, Y. Yao, J. Hu and Z. Deng, "An Online Calibration Method for Robust Multi-Modality 3D Object Detection," *2024 IEEE 11th International Conference on Data Science and Advanced Analytics (DSAA)*, San Diego, CA, USA, pp. 1–10, 2024.

B65. P. Büchel, M. Kratochwil, M. Nagl and D. Rösch, "Deep Calibration of Financial Models: Turning Theory into Practice," *Review of Derivatives Research*, vol. 25, pp. 109–136, 2022.

B66. R. Culkin and S. R. Das, "Machine Learning in Finance: The Case of Deep Learning for Option Pricing," *Journal of Investment Management*, vol. 15, no. 4, pp. 92–100. 2017.

B67. J. Ruf and W. Weiguan, "Neural Networks for Option Pricing and Hedging: A Literature Review," *Journal of Computational Finance*, vol. 24, no. 1, pp. 1–46, 2020.

B68. D. Chaffey and F. Ellis-Chadwick, *Digital Marketing: Strategy, Implementation of and Practice* (5th ed.), Harlow: Pearson Education, 2012.

B69. J. A. Farrell, F. O. Silva, F. Rahman and J. Wendel, "Inertial Measurement Unit Error Modeling Tutorial: Inertial Navigation System State Estimation with Real-Time Sensor Calibration," *IEEE Control Systems Magazine*, vol. 42, no. 6, pp. 40–66, Dec. 2022.

B70. S. S. L. Yang, B. H. S. Lam and C. M. N. Ng, "Digital Sampling Technique in the Calibration of Medical Testing Equipment with Arbitrary Waveforms," *2018 IEEE International Symposium on Medical Measurements and Applications (MeMeA)*, Rome, Italy, pp. 1–6, 2018.

B71. A. Arsenovic et al., "scikit-rf: An Open-Source Python Package for Microwave Network Creation, Analysis, and Calibration [Speaker's Corner]," *IEEE Microwave Magazine*, vol. 23, no. 1, pp. 98–105, Jan. 2022.

B72. D. Belver et al., "ProtoDUNE-DP Light Acquisition and Calibration Software," *IEEE Transactions on Nuclear Science*, vol. 68, no. 9, pp. 2334–2341, Sept. 2021.

B73. Reftab – Equipment Maintenance and Work Order Software https://www.reftab.com/maintenance-management-and-work-order-software (Accessed on 19 April 2025).

B74. Calibration Lab management software: Efficiently manage your calibration data and generate calibration certificates, https://www.metquay.com/calibration-management-software (Accessed on 19 April 2025).

B75. Calibration Software | Calibration System | QT9 QMS https://qt9software.com/qms/calibration-software (Accessed on 19 April 2025).

B76. Y. Shi, D. Kong, X. Ma and C. Zhang, "Dynamic Calibration Method of Blast Pressure Pencil Probes Based on Adjustable Shock Tube," *IEEE Sensors Journal*, vol. 23, no. 11, pp. 11704–11712, June 2023.

B77. Y. Cui, N. Zhou, H. Ouyang, Z. Zou, Z. Gao and W. Tang, "Field Calibration Device and Method of Photogrammetric System for Static Strength Test of Aircraft Structure," *2024 Academic Conference of China Instrument and Control Society (ACCIS)*, Chengdu, China, pp. 78–81, 2024.

B78. S. A. Wani, R. Sarathi and V. Subramanian, "A Multivariable Sensing System for Condition Monitoring of Oil-Immersed Transformers," *IEEE Sensors Journal*, vol. 25, no. 6, pp. 9708–9717, March 2025.

B79. M. Kawalec, K. Czerwi-ska and A. Pacana, "Influence of Technical Condition of Control and Measurement Equipment on Calibration Results," CzOTO 2021, vol. 3, no. 1, pp. 79–88, March 2021.

B80. I. Boureima, V. Gyrya, J. A. Saenz, S. Kurien and M. Francois, "Dynamic Calibration of Differential Equations Using Machine Learning, with Application to Turbulence Models," *Journal of Computational Physics*, vol. 457, Article 110924, May 2022.

B81. A. Nugroho, A. B. Gumelar, E. M. Yuniarno and M. H. Purnomo, "Accelerometer Calibration Method Based on Polynomial Curve Fitting," *2020 International Seminar on Application for Technology of Information and Communication (iSemantic)*, Semarang, Indonesia, pp. 592–596, 2020.

B82. F. Hong, B. Xin, Z. Xin, L. Ke and Z. Haining, "A Design Method for Automatic Calibration System Based on Dynamic Loading and Dynamic Scheduling," *IEEE 2011 10th International Conference on Electronic Measurement & Instruments*, Chengdu, pp. 166–170, 2011.

B83. P. Sun, F. Zhou, L. Wang, W. Zhang, Y. Liu and L. Yang, "A Novel Global Camera Calibration Method Based on Flexible Multidirectional Target," *IEEE Transactions on Instrumentation and Measurement*, vol. 72, pp. 1–13, 2023.

B84. T. Nelson et al., "The SuperCam Instrument for the Mars 2020 Rover," *2020 IEEE Aerospace Conference*, Big Sky, MT, USA, pp. 1–12, 2020.

B85. H. C. Miles, M. D. Gunn and A. J. Coates, "Seeing through the "Science Eyes" of the ExoMars Rover," *IEEE Computer Graphics and Applications*, vol. 40, no. 2, pp. 71–81, March-April 2020.

B86. N. U. Okafor and D. T. Delaney, "Missing Data Imputation on IoT Sensor Networks: Implications for On-Site Sensor Calibration," *IEEE Sensors Journal*, vol. 21, no. 20, pp. 22833–22845, Oct. 2021.

B87. L. Yuxin, L. Wen, W. Dongyan, J. Xinchun and S. Ge, "Online Calibration Method of Smartphone Magnetometer in Vehicle Geomagnetic Matching Positioning," *2022 IEEE 12th International Conference on Indoor Positioning and Indoor Navigation (IPIN)*, Beijing, China, pp. 1–8, 2022.

B88. J. Hai-lin, W. Shu-wen, L. Hui and X. Dong-lei, "Theoretical Investigation on the Line-of-Sight Measurement Error of the on-Orbit Geometric Calibration of the Geo-Stationary Earth Orbit Satellite," *2015 International Conference on Optoelectronics and Microelectronics (ICOM)*, Changchun, China, pp. 227–229. 2015.

B89. P. Kangaslahti, A. Tanner, W. Wilson, S. Dinardo and B. Lambrigsten, "Prototype Development of a Geo Stationary Synthetic Thinned Aperture Radiometer (GeoSTAR)," *IEEE MTT-S International Microwave Symposium Digest, 2005*, Long Beach, CA, USA, pp. 2075–2078, 2005.

B90. W. Bin et al., "Key Technology Research of Satellite-Ground Integrated Design on Optical Remote Sensing Satellite," *2024 IEEE International Conference on Control Science and Systems Engineering (ICCSSE)*, Beijing, China, pp. 475–479, 2024.

3 Mathematical Methods in Calibration

3.1 INTRODUCTION

Mathematical descriptions are necessary in every stages of measurements and calibrations. Involvement of mathematics starts with the description of physical principles and models of measurands to the final stages of evaluations and decision making in calibrations. Popular mathematical methods used in calibration can broadly be listed as follows:

Physical models
Empirical models
Statistical and stochastic models
Data-based calibration models
Ground truth-based calibration models

A combination of methods and many other descriptions depending on the characteristics of measurands and methods of analyses is employed.

In the following subsections, mathematical principles for calibrations will be explained.

3.2 ALGORITHMS FOR CALIBRATIONS

Algorithms in calibrations are used for many reasons, including to create best-fit calibration curves to capture the relationship between the dependent and independent variables. The simplest but frequently encountered curve is the linear calibration curve. However, most software allows for the choice of multiple different curve fittings using optimal curve fitting rules. Optimality occurs when the correlation coefficient of fit is closest to the value of 1.0.

Calibration algorithms provide the following calibration curves fit:

1. Point-to-point fit is the intermediate curve fit between points.
2. Linear fit is the least squares line of the best fit.
3. Linear origin fit is the least-squares line of best fit forced through the origin.
4. Quadratic fit is the second-order least squares of best fit.
5. Quadratic origin fit is the second-order least-squares curve of best fit forced through zero.
6. Cubic fit is the third-order least-squares curve of best fit.
7. Cubic origin fit is the third-order least-squares curve of best fit forced through zero.

DOI: 10.1201/9781003590767-3

8. Higher order fit is the higher-order least-squares curve of best fit.
9. High-order origin fit is the high-order least squares curve of best fit through zero.
10. Logarithmic fit is the logarithmic least-squares curve of best fit.
11. Exponential fit is the exponential least squares curve of best fit.
12. Statistical and stochastic curve fit.
13. Other mathematical fits provide multidimensional order least squares curve of best fit. Examples of such fits are ordinary differential equations (ODE) or partial differential equations of various degrees.

The calibration algorithms must be robust and suitable for the calibration in hand. These algorithms need assessments to determine a balance between model complexity and the available sample size. Poor algorithms can be misleading and can potentially be harmful.[C1]

Solutions for calibration require two fundamental factors: (1) mathematical descriptions and (2) calibration charts and curves of calibration. Professional off-the-shelf algorithms are available for curve fit, including X-Rite, Imatest, or specific software like Fiji, and many others.[C2]

The majority of calibration algorithms are based on approximate linear calibration models and low-order polynomials. If stochastic techniques are involved, the estimates of the calibration line parameters, uncertainties, coverage intervals, and the associated probability distributions can be obtained using various methods such as Monte Carlo techniques, and many others.[C3]

Calibration of parametric ODE and partial differential equations (PDE) is an evolutionary model against ground truth data that take advantage of recent developments in AI. Various forms of software are available off-the-shelf for constructing custom AI models.[C4]

AI algorithms are used in calibrations extensively, including alternative solutions for difficult and complex mathematical descriptions. For instance, back-propagation neural network for representing PDEs can be trained for solving complex PDEs. Using variational formulations, the solution problem can be expressed in terms of the minimization of a particular integral. Then, the problem can be solved using gradient methods. AI libraries such as Armadillo, Matplotlib, OpenNN, JAX, PyTorch, SciPy, TensorFlow, and others enable the determination of appropriate regression, classification and clustering for effective curve fitting.[B80] [C5]

3.3 CHARTS AND CURVES

Measuring devices are characterized by a specific response function, which may be linear, polynomial, logarithmic, exponential, ODE, PDE, stochastic, or in other mathematical forms. This depends on the nature and behavior of the measuring device, physics, system being measured, process of measuring, and the process itself. The data from the input and output relationships are plotted to describe the relationship are calibration curves and charts.

A *calibration curve (standard curve)* is a way of determining the input/output relationship of a device by comparing it with a set of standard samples of known values.

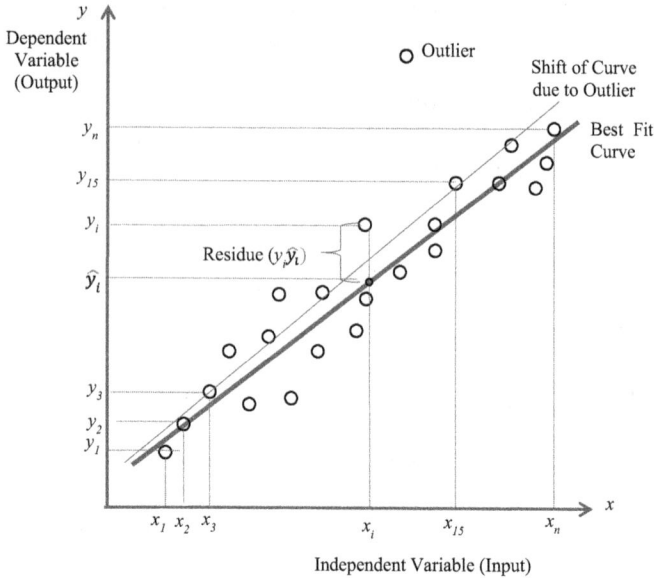

FIGURE 3.1 A linear regression line. Many measurement devices are likely to have a linear relationship between the input and output variables. Or at least their output is linearized by selecting and using suitable electronic components. Linear relationships lead to relatively easy and inexpensive calibrations. Linear regression analysis is one of the key methods in calibration applications of AI models.

A calibration curve is a plot of how the response of a device changes with the variable under measurement.[C6] Calibration curves can be linear or nonlinear (e.g., polynomial) or stochastic. The linear best-fit curves will be explained in detail next. Other curve fits demonstrate similar mathematical foundations based on different mathematical descriptions.

Linear regression finds the line that best (Figure 3.1) fits the independent data points x_i and dependent data points y_i, which can be expressed as:

$$y = ax + b \tag{3.3.1}$$

Where a is the slope of the line and b is the intercept. They can be calculated from the data points such as:

$$b = \frac{\left[n\left(\sum y \sum x^2\right) - \left(\sum x\right)\left(\sum xy\right) \right]}{\left[n\left(\sum x^2\right) - \left(\sum x\right)^2 \right]} \tag{3.3.2}$$

$$a = m = \frac{\left[n\left(\sum xy\right) - \left(\sum x\right)\left(\sum y\right) \right]}{\left[n\left(\sum x^2\right) - \left(\sum x\right)^2 \right]} \tag{3.3.3}$$

Where x is the independent variable, y is the dependent variable, a (m) is the slope of the line, and b is the intercept.

Calibration results in the curves are tested by the *goodness of fit indicators* such as residual standard deviation, mean absolute error, root mean square error, correlation coefficient, and coefficient of determination (R^2).[C7]

The formulae describing the goodness of fit indicators are:

Mean absolute error

$$\text{MAE} = \frac{1}{n} \sum_{i=1}^{n} |y_i - \hat{y}_i| \tag{3.3.4}$$

Where n is the number of samples, y_i is the measured value, and $\hat{y}_i$ is the predicted value of the ith sample.

Mean square error

$$\text{MSE} = \frac{1}{n} \sum_{i=1}^{n} (y_i - \hat{y}_i)^2 \tag{3.3.5}$$

Where n is the number of samples, y_i is the measured value, and $\hat{y}_i$ is the predicted value of the ith sample.

Residual standard deviation

$$S(r) = \sqrt{\frac{\sum_{i=1}^{n} (y_i - \hat{y}_i)^2}{n-2}} \tag{3.3.6}$$

Where $(\hat{y}_i)$ is the value of y_i predicted by the equation on the calibration line for a given value of x_i.

Root mean square error

$$\text{RMSE} = \sqrt{\frac{\sum_{i=1}^{n} (y_i - \hat{y}_i)^2}{n}} \tag{3.3.7}$$

Coefficient of determination

$$R^2 = 1 - \frac{\text{SSR}}{\text{SST}} = 1 - \frac{\sum_{i=1}^{n} (\bar{y}_i - \hat{y}_i)^2}{\sum_{i=1}^{n} (y_i - \hat{y}_i)^2} \tag{3.3.8}$$

Where SSR is the some of the squared regression, SST is the total variation, y_i is the y value of the observation i, $\bar{y}$ is the mean of the y value, and $\hat{y}$ is the predicted value of observation i.

Correlation coefficient

$$r = \frac{n\sum (xy) - \left(\sum x\right)\left(\sum y\right)}{\sqrt{\left[n\sum x^2 - \left(\sum x\right)^2\right]\left[n\sum y^2 - \left(\sum y\right)^2\right]}} \qquad (3.3.9)$$

For all x and y values

The straight line through calibration data obtained from the equation of the best-fit straight line and correlation coefficient, r. There are various correlation coefficients, and the one often used is the Pearson correlation coefficient given in Equation 3.3.9. The Pearson r value provides a measure of the degree to which the values of x and y are linearly correlated.

Device response data are normally plotted on the y-axis and the standard value (reference value) on the x-axis. This is because the statistics used in the regression analysis assume that the errors in the values on the x-axis are insignificant compared with those on the y-axis. This assumption may not be true in many cases and often is not.

Most measurements exhibit linear relationships leading to well-established regression analysis. Linear regression requires a good understanding of the concept of residuals that may bear similarities in other complex curve fits (e.g., polynomial). A *residual* is the difference between an observed y value and the value calculated using the equation of the best-fit line (Figure 3.1). The residuals provide an indication of how well the line fits the data. The sum of the squared residuals for a poorly fit will be much larger than in the best-fit line. It can be shown that the line that gives the smallest sum of the squared residuals represents best of the linear relationship between the x and y variables. Software for linear regression simply calculates the values for a and b that minimize (or optimize from a set of repeated calculations) the sum of the squared residuals. This type of regression is referred to as "the least squares regression".[C8]

Determining the best-fit curves will have pitfalls due to many reasons, and one should be aware of some pitfalls in determining the calibration curves and charts, given as follows:

- Calibration standards do not adequately cover the intended range of measurements.
- Calibration results are not evenly spaced across the calibration range.
- Uncertainty associated with the calibration is too large.
- Wrong regression method is implemented.
- The calibration line is fitted through zero when the actual interception differs from zero.
- Instrument software is used to carry out regression is suitable, but the plot of the calibration data is not complete.
- Residual standard deviation is used as an estimate of the uncertainty, rather than carrying out the full standard error of prediction.

- The instrument used to make measurements is not within specification.
- There are too many outliers in the measurements.

Outliers are generated inevitably during the process of calibration. Outliers are points of influence that introduce errors in the readings and cause biases in the position of the curve. An outlier will shift or distort the best regression line up or down, and the gradient of the best fit will be altered.

There are a host of algorithms to process calibration data such as Microsoft Excel, LabVIEW, and many others. Microsoft Excel, for example, can be used to generate different charts, which can be compared with other charts generated by customized calibration software.

3.4 DATA HANDLING AND DATA QUALITY

Sensor-based devices are basic building blocks of observing physical phenomena. These devices communicate with other devices in a digital medium. The data produced are transferred via various communication networks. Decisions at the receiving end are made based on information coming from the devices. Measurement systems and sensors use wireless communication protocols such as Wi-Fi, Bluetooth, Zigbee, or Z-Wave to transmit data. Sensor faults and failures can severely erode data quality.

Device faults and failures can severely erode data quality. When a device generates false information or fails, the transferred data become corrupt, thus giving erroneous and conflicting information. This may result in a compromise of the performance of the overall system. Device failures can be detected by three methods: model-based or signature-based or data-driven methods. The *model-based methods* use mathematical models of the system to estimate the true values by comparison of predicted and measured values. The *signature-based methods* compare current device readings with pre-defined threshold values. The *data-driven methods* analyze the statistical properties of the device data to learn the underlying patterns and relationships. Four typical AI models used in fault and failure detection are decision tree, k-nearest neighbor, Gaussian Naïve Bayes, and random forest.[C9]

To ensure quality and confidence in any calibration, data analysis becomes an important factor, and it has several phases, as shown in Figure 3.2. The essential steps in data analysis are collection of relevant data, interpretation of data, pre-processing, initial analysis, model selection and application, and assessing the relevance of results. Calibration data can be analyzed by descriptive, diagnostic, predictive, and prescriptive methods. A couple of tools that can be used in data analysis are regression methods, correlations, hypothesis testing, statistical techniques, and AI models.

Depending on the application, the calibration data can range from only a few data points in some cases to a massive amount of data in others. In all cases, data were examined to identify patterns, trends, and correlations. Data generated by measuring devices are handled in two basic ways: by using parametric methods or non-parametric methods. *Parametric methods* are highly effective in simple systems, but in many applications, they cannot fully capture complex cases. *Non-parametric methods* use statistical techniques, regression methods, and AI models. AI models have proven to be effective in complex cases, but they come with increased computational costs.[C10]

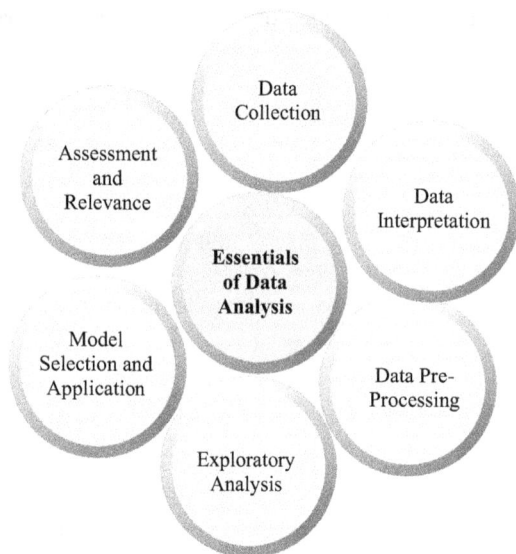

FIGURE 3.2 Essential components of data analysis. Data preparation and analysis play an important role in the successful implementation of calibration in both traditional and AI applications. Data needs to be preprocessed, and the relevant data must be selected carefully.

Calibration datasets should contain all the components of the expected variables in the representative data. The *representative dataset* implies a dataset that incorporates all sources of possible known and unknown variables and variations. The sample range for a successful calibration should exceed the range of variables expected in the samples. In addition, it is necessary to examine and identify the outliers or other irregularities by preprocessing the data. Therefore, selecting a representative dataset for building a calibration model is a delicate step.

The selection of a representative calibration dataset can be done in two basic ways: by mathematical algorithms or clustering analysis.

Mathematical algorithms are largely based on statistical principles and selected sample sets. Selected sample sets may not be the optimal subset of the overall data. Several tests may be needed to determine the representativeness of the selected data with respect to the total set.

Calibration data selection becomes more difficult as variations in data increase. Among many others, algorithms such as the Kennard and Stone algorithm and Federov algorithm, using the D-optimality criterion, can be used to select an optimal sample subset from the measurements. The Kennard and Stone algorithm randomly splits the dataset into a calibration dataset and a validation dataset. The calibration dataset is used to fit the model and validation dataset checks that trained model works on a set of unknown data. The Federov algorithm is based on spreading data on various servers and devices for the purpose of privacy protection and security of data. For example, the Federov algorithm selects calibration samples that optimally span the domain of interest based on the optimality criterion *a priori*. The Duplex algorithm, a variant of the Kennard–Stone algorithm, is an alternative for selection of the calibration and test datasets.

The Kennard–Stone algorithm is used in uniform distributions over the predictor space. It begins by selecting the pair of points that are furthest apart from each other. [C11] The Federov algorithm addresses the optimal design of regression models where defining a valid regression model is essential. There are several improved versions, such as the Modified Federov algorithms and the K-exchange algorithm. [C12]

Cluster analysis selects the samples that are the furthest from the center of each cluster; the procedure is iterated until the desired number of samples is selected. In this way, the extremes are covered but not necessarily at the center of the data, thus leading to an inhomogeneous distribution throughout the data range.

Algorithms are developed to choose the most representative samples of the experimental domain for building calibration models. They consider all the variability of the calibration space using the variance of the samples or their distance from each other. Consequently, the quality and robustness of the model are improved by considering the best calibration set that spans the entire experimental domain, including the extreme points and any out-of-range calibration points. [C13]

3.4.1 DATA QUALITY

Data quality is an indication of the degree of accuracy, consistency, completeness, reliability, and relevance of data. High-quality data leads to accurate analyses, implementation of effective strategies, and well-informed decisions. Data quality can be influenced by data collection methods, process of data entry, data storage, and data integration. The European Union Data Quality Directive (EU-DQD) defines the Data Quality Objective (DQO) as monitoring methods that must comply with indicative measurement for regulatory purposes. The EU-DQD is a measure of acceptable uncertainty for indicative measurements. [C14]

Quality of data in calibrations is significant, particularly in health and medicine, robotics, space exploration, research and design, and national defense applications. Data need to be valid, accurate, consistent, complete, reliable, and well-structured (Figure 3.3).

Low quality of data reduces performance if the data are incomplete and contain missing data and significant errors. A deep learning neural network (e.g., Jordan network) can be used as a model to predict the incomplete and missing data. [C15]

3.4.2 MISSING DATA

Missing data affect most real-world datasets, eroding the data quality in critical applications such as medical records, geo-informatics, monitoring traffic flow, and numerous industrial applications. Two common methods for handling missing data are downsampling and imputation. *Downsampling* discards an incomplete observation by dropping or disregarding missing data records. Downsampling is a simple and effective method, but it can result in losing useful information from valuable data. *Imputation* replaces missing values by using either single imputation or multiple imputation. Single imputation replaces all missing values with a single value, such as a value of zero or the mean of the observation. Imputing missing data uses various

FIGURE 3.3 Data quality essentials. Data quality is essential in large-scale applications. Reliable and good quality data need to be accurate, consistent, and complete. A well-structured valid data ensures successful calibration.

methods such as listwise deletion, pairwise deletion, and model-based imputation. A hybrid algorithm can be used by dividing the dataset into two categories in complete and incomplete datasets. At each iteration, the algorithm trains the neural network to predict a specific missing value. Multiple imputation is an iterative model using discriminative and generative methods. Discriminative methods include denoising autoencoders, random forest-based imputation (Missforest), and matrix completion. Generative methods consist mostly of techniques such as variational auto encoders, K-nearest neighbor, and generative adversarial network.[B86] [C16] [C17]

3.5   MIXUP CALIBRATIONS

Mixup is a data augmentation that generates a weighted combination of random pairs from the training data. Mixup improves calibration across various model architectures and datasets. It can implicitly perform label smoothing; hence, overconfidence issues can be minimized or totally avoided.

Models trained with mixup have been shown to perform well on uncertainty in calibration. This is an important consideration in many real-world applications. Mixup's calibration problem by considering the training stage and post-hoc processing as a unified system.[C18] [C19]

By taking convex combinations between pairs of samples and their labels, mixup training has been shown to improve predictive accuracy. By decomposing the mixup process into data transformation and random perturbation, the confidence penalty nature of the data transformation calibration degradation can be avoided. Mixup inference in training adopts a simple decoupling principle for recovering the outputs of raw samples at the end of the forward network pass. By embedding mixup inference, models can be learned from the original hot labels and hence avoid the negative impact of the confidence penalty.[C20]

3.6 CALIBRATION PROCESS AND HANDLING ERRORS

Calibration intends to eliminate errors and reduce biases over a continuous range of values. For this, a functional relationship must be established between the reference standard and the corresponding measurements.[A2] [C21]

A calibration procedure for error reduction should have the following steps:

- Selection of an appropriate reference standard with known values covering the entire range of interest.
- Conducting the calibration procedure within the requirements of the reference standard.
- Establishing the calibration curves (e.g., least-squares fit) to observe the relationship between the measured values and the values of the reference standard.
- Correction of measurements by adjusting the necessary parameters or by using calibration curves. This can be hardware adjustment when necessary.
- Adhering to the traceability requirements.
- Validation of calibration results. Firm knowledge of the realms of the calibration process must be worked out for validation.
- Preparation of the appropriate documentation of the calibration procedure, results analysis, and interpretation of results.
- Careful planning and revision of plans to improve the calibration process.[A2]

During the calibration, the readings of the test item are compared with the ideal values or reference standards. In many situations, it may not be possible to achieve a perfect calibration because of measurement biases and uncontrollable random errors, which may be expressed as:

$$\text{Ideal value} = \text{Measured value} + \text{Bias} + \text{Error}$$

Similarly, the reference values may be subject to bias and error.

$$\text{Ideal reference value} = \text{Reference value} + \text{Bias} + \text{Error}$$

Assuming the bias errors are the same for the ideal value and ideal reference value, the deficiency can be expressed as

$$\text{Deficiency} = \text{Ideal measured value} - \text{Ideal reference value}$$

Due to the randomness of errors, this deficiency may not be zero; therefore, the calibration measurements may have to be repeated many times. Then, suitable statistical techniques can be applied to determine the calibration curves, average readings, uncertainties, and standard deviations. The process of collecting data for creating the calibration curve is critical in a successful calibration program.

The deficiency between the ideal and observed values determines accuracy. Every measurement contains some known, some unknown, and some unknowable inaccuracies. Measurement uncertainty is the result of unknown inaccuracies.

Calibrations encounter systematic errors and random errors. Systematic errors tend to be constant and repeatable due to faulty equipment and incorrect calibration procedures. If the error source is not known, systematic errors may be estimated from the mean values of a series of repeated measurements. Random errors describe the scatter of repeated measurements around the mean value. Random errors are always present in measurements.[C22]

Traceable measurement requires an unbroken chain of comparisons back to a reference value within acceptable errors within a reasonable uncertainty band shown. Measurement conditions determine the calibration validity. The distinction between the calibration validity conditions and the conditions of subsequent measurements determines the validity of traceability.

Often, the error of a measurement result is never known exactly since the value of a measurand can never be the same. However, useful estimates of an error are possible when the uncertainty in the error is small relative to the magnitude of the error. Hence, errors can only be estimated by performing measurements and comparing the results with a standard that has some previously assigned values. This assigned value can be from the fundamental standards or estimate of the "true value" of the measurand.

Most calibrations are conducted with the aid of computers to capture and analyze data. Once the results are obtained, software packages assist in the analysis of the information. Most packages use the method of least squares for estimating the coefficients. Some of the packages can perform a weighted fit if the errors of the measurements are not constant over the calibration interval. The software packages provide information such as the coefficient of the calibration curve, standard deviations, residual standard deviation of the fit, and goodness of the fit.

3.7   UNCERTAINTY IN CALIBRATIONS

The ISO International Vocabulary of Basic and General Terms in Metrology (VIM) defines uncertainty as "a parameter, associated with the result of a measurement, that characterizes the dispersion of the values that could be reasonably attributed to the measurand." Measurement uncertainty assessment summarizes the combined effects of all uncertainty sources in terms of a single quantity. This is called "the combined standard uncertainty". The ISO guidelines, such as ISO 11095, give guidance on how to assess, correct, and calculate uncertainties.[A1]

Uncertainties are expressed within uncertainty budgets. The uncertainty budget is obtained by adding uncertainties of all known affecting components together. In the uncertainty budget, often, the Test Uncertainty Ratio (TUR) of "4 to 1" has been implemented. It basically says that an appropriate uncertainty relationship between the calibration standard and the device under test (DUT) should be at 4 to 1, meaning the uncertainty of the reference measurement is four times smaller than the uncertainty of the DUT.[C23]

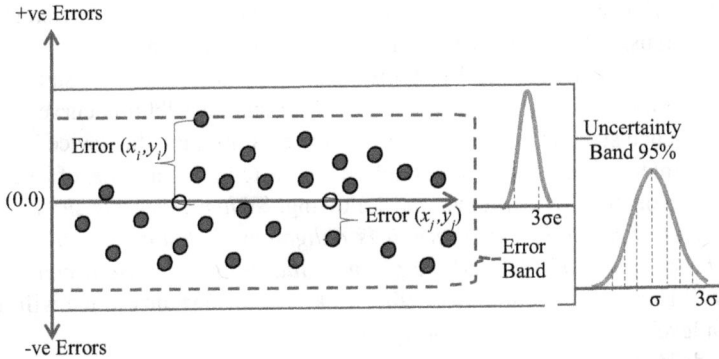

FIGURE 3.4 Errors in calibration. Errors in measurements occur because of variations in measurand, response of the sensing element, processing of signals, transmission of data, and variations in data handling. Positive and negative errors can be plotted from ground truth measurements and the expected (predicted) values. Errors should be bound within the uncertainty band with a certainty (say 95%) that can be attributed to the measurements. Error distribution can mostly be described by a Gaussian distribution, but it may not be the case, particularly in highly nonlinear measurements.

Uncertainty is defined as "the dispersion of values that can be attributed to the measurand." For example, a 95% level of confidence implies that 95% of the values can reasonably be attributed to the measurand within an uncertainty band. Similarly, a 99% level of confidence implies that 99% of the values can be attributed to the measurand within an uncertainty band (Figure 3.4).

Some of the sources of uncertainty are the shortfalls within the measurement standards selected, the device under calibration, and the calibration technique employed. Evaluation of uncertainty requires multiple components, which are classified as Type A uncertainty and Type B uncertainty. Type A uses statistical procedures, such as average and standard deviation. Type B is evaluated based on scientific judgement, previous measurement data, manufacturer's specifications, data provided in calibration certificates, and so on.[C24]

Common sources of measurement uncertainty are instrument accuracy, environmental conditions, human error, calibration standards, and repeatability and reproducibility. *Instrument accuracy* depends on the precision of the measuring instrument. Precision plays a significant role in uncertainty. Devices with higher accuracy levels tend to have lower uncertainty. *Environmental conditions* such as the variations in temperature, humidity, and pressure affect measurements. Calibrating instruments in controlled environments reduces sources of uncertainty. *Human errors* are the mistakes made during measurements, such as the parallax errors or the application of incorrect technique contribute to uncertainty. *Calibration standards* indicate the accuracy and traceability of standards used in the calibration process, thus influencing measurement uncertainty. Calibrating instruments against standards with higher degrees of uncertainty propagates uncertainty to the measured results. *Repeatability and reproducibility* are the variations in the measurements taken by different operators, although using the same instrument can introduce uncertainty.

Most uncertainty results tend to be toward the middle of the possible range of measurement data, thus largely following the Gaussian distribution curve. The calibration reference standard should have lower uncertainty than the device being calibrated. The international metrology standard, Guide to the Expression of Uncertainty in Measurement (GUM) expresses the following. *"When reporting the result of a measurement of a physical quantity, it is obligatory that some quantitative indication of the quality of the result be given so that those who use it can assess its reliability."* Hence, a measurement without a known uncertainty value will have an unknown level of quality and reliability.[C25]

Methods for reducing uncertainty measurement are calibration, correct operations, environmental controls, repeatability studies, uncertainty analysis, instrument maintenance, and using multiple instruments. Regular *calibration* of measuring devices against standards with known traceable accuracies is a fundamental method to reduce uncertainty. A comprehensive uncertainty analysis involves identifying and quantifying all sources of errors in a measurement process. By understanding where uncertainty originates from, the corrective steps can be taken. When high accuracy is essential, using multiple instruments and cross-checking measurements with different methods provides robust and accurate measurements, thereby reducing uncertainty.[C26] [C27]

Classical uncertainty estimations mainly focus on probabilistic methods. However, the determination of calibration uncertainty by AI models attracted considerable interest in recent years. It is important to understand that uncertainty does not mean accuracy or inaccuracy. Accuracy describes the measurement capability of a device, while uncertainty describes the measurement specification.[C28]

Traditional uncertainty estimation approaches mainly rely on sampling methods, such as Monte Carlo (MC) sampling. MC is an effective method to approximate an exact posterior inference, which has been a popular method for uncertainty estimation. However, it becomes a slow and computationally expensive process when integrated into a deep learning architecture.[C29]

Uncertainty estimates can also be made by employing AI models. They identify out-of-distribution samples, detect anomalies/outliers, delegate high-risk predictions, defend against adversarial attacks, and so forth. Calibration error is commonly adopted for evaluating the quality of uncertainty estimators in deep neural networks. Such a metric is beneficial for training predictive models, even though uncertainties may not be explicitly targeted.[C30]

3.8　FACTORS IN CALIBRATION

Calibration factors allow for comparison of device output to known standards. Calibration factors are numerical values that multiply the raw output of a device for accurate measurements in a different unit. For example, calibration factors are used if a medium for temperature readings is required, say in Kelvins, to a different scale that the device has been calibrated for, say in Celsius. The calibration factor helps to increase the accuracy of the measurement.[C31]

3.9 INTERVALS IN CALIBRATION

The calibration interval is the number of days between scheduled calibrations. It is also expressed as calibration frequency, calibration period, or calibration due date. The user of a measuring device should determine the calibration intervals depending on the frequency of use. Determination of the long-term calibration plan and calibration interval is the responsibility of the in-house management. This is due to legal requirements, product consistency, safety, consistency of reading, marketing, and other reasons. Calibration intervals should clearly be worked out, effectively implemented, and properly documented.

Determining the calibration intervals is an essential part of maintenance and quality control. Calibration intervals have two main steps: (1) establish an initial interval and (2) determine the fixed calibration interval thereafter. Appropriate data collected from the devices to determine intervals.[C32]

Suggesting calibration intervals to clients by calibration laboratories is not recommended by ISO 17025 standard, clause 7.8.4.3. It states that *"A calibration certificate or calibration label shall not contain any recommendation on the calibration interval, except where this has been agreed with the customer."* The main reason for this is that once the instrument is out of the laboratory, the lab does not have control of the way the device is used and the frequency of usage. Therefore, compliance with the recommended interval cannot be guaranteed.[A9]

3.10 SAMPLING ISSUES

The accuracy of calibration is influenced by the process of sampling and the selection of sampling techniques. Appropriately worked out sampling methods are essential for successful calibration and analysis of its results. Sampling data infer important information about a larger population from a smaller group of data. Sampling can broadly be categorized as probability sampling and nonprobability sampling.[C33]

3.10.1 PROBABILITY SAMPLING

Probability sampling for calibration can be classified as random, stratified, systematic, cluster, and multi-stage sampling. *Random sampling* is a method of selecting a sample from a population in which each member of the population has an equal probability of being selected. *Stratified random sampling* divides a population into subgroups or strata based on certain characteristics for samples to be selected randomly from each stratum. *Systematic sampling* involves selecting a sample from a larger population using a predetermined interval. *Cluster sampling* requires partitioning a population into clusters, from which a random sample of members is selected. *Multi-stage sampling* is a variant of cluster sampling that involves the selection of samples in two or more stages.[C34]

3.10.2 NONPROBABILITY SAMPLING

Nonprobability sampling is a technique in which the likelihood of each member of the population being selected is not known. Nonprobability sampling can be purposeful sampling, snowball sampling, and quota sampling. In *purposive sampling or deliberate sampling*, individual samples are selected based on their relevance to the objectives. *Snowball sampling* is a chain sampling or sequential sampling when a sample recruits additional related samples such as selecting a friend or a relative. *Quota sampling* picks samples based on a set of predetermined criteria.[C35]

Sampling methods define the types of calibration methods required. An identified calibration method involves formulating a hypothesis, suggesting solutions, collecting and organizing samples, evaluating the data, making deductions, and reaching conclusions. The results are tested and analyzed to determine if the method selected leads to representative samples and the formulation of a hypothesis.

Calibration AI model building is composed of a suitable number of samples and the selection of suitable regression techniques. In the multivariate regression, the sample numbers and sample collection must ensure adequate representation of the target population. Sample selection and sample number are the cornerstone of the robustness of applied models. Correct sampling and sampling accuracy ensure a high confidence level in results, reduction in relative standard errors, and desired representation of the data in an application. The required sampling number can be calculated by the statistical parameters of the population and robust representation. Determination of relative standard error is an important influencing factor related to the statistical parameters of a sample set.[C36] [C37]

3.11   CONCLUSIONS

Mathematics plays an important role in calibrations. Mathematical descriptions of processes lead the way to calibration analysis and regressions. The results can be evaluated, and accuracy and uncertainties can be calculated to have a good idea about the behavior of a system. Correlation coefficients, root mean square analysis, probability curves, and other testing methods are valid in AI applications as much as the classical approaches.

REFERENCES

C1. B. Van Calster, D. J. Mclernon, M. van Smeden and L. Wynants, "Calibration: The Achilles Heel of Predictive Analytics," *BMC Medicine,* vol. 17, p. 230, 2019.

C2. M. A. Barbero-Álvarez, J. A. Rodrigo and J. M. Menéndez, "Self-Designed Color Chart and a Multi-Dimensional Calibration Approach for Cultural Heritage Preventive Preservation," *IEEE Access*, vol. 9, pp. 138371–138384, 2021.

C3. J. Palencár, R. Palencár, M. Chytil, G. Wimmer, G. Wimmer and V. Witkovský, "ISO Linear Calibration and Measurement Uncertainty of the Result Obtained with the Calibrated Instrument," *Measurement Science Review*, vol. 22, no. 6, pp. 293–307, 2022.

C4. J. Miguez, H. Molina-Bulla and I. P. Mariño, "A Sequential Monte Carlo Method for Parameter Estimation in Nonlinear Stochastic PDE's with Periodic Boundary Conditions," *2023 IEEE 9th International Workshop on Computational Advances in Multi-Sensor Adaptive Processing (CAMSAP)*, Herradura, Costa Rica, pp. 86–90, 2023.

C5. A. Bhadwal, *15 Best Machine Learning Libraries You Should Know in 2025,* 30 Jan. 2025. https://hackr.io/blog/best-machine-learning-libraries (Accessed on 19 April 2025).

C6. P. W. Chen, N. Wary, L. Wang, Q. Wang and A. C. Carusone, "All-Digital Calibration Algorithms to Correct for Static Non-Linearities in ADCs," *2020 IEEE International Symposium on Circuits and Systems (ISCAS)*, Seville, Spain, pp. 1–5, 2020.

C7. L. Prichard and V. Barwick, *"Preparation of Calibration Curves—A Guide to Best Practice,"* Technical Reports, LGC/VAM/2003/03, pp. 1–27, 2003. https://www.researchgate.net/publication/334063221 (Accessed on 19 April 2025).

C8. D. F. Williams, B. Jamroz and J. D. Rezac, "Evaluating Uncertainty of Nonlinear Microwave Calibration Models with Regression Residuals," *IEEE Transactions on Microwave Theory and Techniques*, vol. 68, no. 9, pp. 3776–3782, Sept. 2020.

C9. E. A. Hinojosa-Palafox, O. M. Rodríguez-Elías, J. H. Pacheco-Ramírez, J. A. Hoyo-Montaño, M. Pérez-Patricio and D. F. Espejel-Blanco, "A Novel Unsupervised Anomaly Detection Framework for Early Fault Detection in Complex Industrial Settings," *IEEE Access*, vol. 12, pp. 181823–181845, 2024.

C10. J. Springer, M. Oispuu, W. Koch and P. Knott, "Array Calibration Using Neural Networks," *2023 IEEE 9th International Workshop on Computational Advances in Multi-Sensor Adaptive Processing (CAMSAP)*, Herradura, Costa Rica, pp. 51–55, 2023.

C11. R. Kennard and L. Stone "Computer Aided Design of Experiments," *Technometrics*, vol. 11, no. 1, pp. 137–148, 1969.

C12. V. V. Fedorov, "Theory of Optimal Experiments (Review)," *Biometrika*, vol. 59, no. 3, 697–698, 1972.

C13. S. Brown, B. Walczak and R. Tauler, *Comprehensive Chemometrics, Chemical and Biochemical Data Analysis* (2nd ed.). Elsevier, London, UK, 2020.

C14. Data Quality Guidelines, Publication Office of the European Union, Nov. 2021. https://op.europa.eu/webpub/op/data-quality-guidelines/en/ (Accessed on 19 April 2025).

C15. N. Al-Milli and W. Almobaideen, "Hybrid Neural Network to Impute Missing Data for IoT Applications," *2019 IEEE Jordan International Joint Conference on Electrical Engineering and Information Technology (JEEIT)*, Amman, Jordan, pp. 121–125, 2019.

C16. M. Pazhoohesh, Z. Pourmirza and S. Walker, "A Comparison of Methods for Missing Data Treatment in Building Sensor Data," *2019 IEEE 7th International Conference on Smart Energy Grid Engineering (SEGE)*, Oshawa, ON, Canada, pp. 255–259, 2019.

C17. B. Agbo, H. Al-Aqrabi, T. Alsboui, M. Hussain and R. Hill, "Missing Data Recovery for Downstream Learning in Low-Cost Environmental Sensor Networks," *2023 9th International Conference on Information Technology Trends (ITT)*, Dubai, United Arab Emirates, pp. 26–31, 2023.

C18. D. B. Wang, L. Li, P. Zhao, P. A. Heng and M. L. Zhang, "On the Pitfall of Mixup for Uncertainty Calibration," *2023 IEEE/CVF Conference on Computer Vision and Pattern Recognition (CVPR)*, Vancouver, BC, Canada, pp. 7609–7618, 2023.

C19. D. Y. Kim, D. K. Han, J. H. Jeong and S. W. Lee, "Calibration-Free Driver Drowsiness Classification with Prototype-Based Multi-Domain Mixup," *IEEE Transactions on Intelligent Transportation Systems*, vol. 26, no. 3, pp. 2955–2966, March 2025.

C20. J. Noh, H. Park, J. Lee and B. Ham, "RankMixup: Ranking-Based Mixup Training for Network Calibration," *2023 IEEE/CVF International Conference on Computer Vision (ICCV)*, Paris, France, pp. 1358–1368, 2023.

C21. V. Kopke, S. C. Mourão and M. Brito, "Determination of the Calibration Interval of Measuring Instruments: Which Method Should I Use?" *IEEE Instrumentation & Measurement Magazine*, vol. 27, no. 2, pp. 59–62, April 2024.

C22. J. Zhao, "Research on the Error Compensation System of Automatic Calibration of Instruments Based on Deep Neural Network," *2024 IEEE 4th International Conference on Data Science and Computer Application (ICDSCA)*, Dalian, China, pp. 550–554, 2024.

C23. Q. Yang, L. Xu, L. Xia, Z. Chen, S. Zhang and R. Zhou, "Performance Ratio Test and Uncertainty Evaluation of Photovoltaic Power Generation System Based on Linear Regression," *2021 IEEE 5th Conference on Energy Internet and Energy System Integration* (EI2), Taiyuan, China, pp. 1911–1961, 2021.

C24. I. Zakharov and O. Botsiura, "Advanced Methods for Measurement Uncertainty Evaluation," *2022 XXXII International Scientific Symposium Metrology and Metrology Assurance (MMA)*, Sozopol, Bulgaria, pp. 1–4, 2022.

C25. ISO/IEC Guide 98-1:2024(en), *Guide to the Expression of Uncertainty in Measurement—Part 1: Introduction.* https://www.bipm.org/documents/20126/2071204/JCGM_GUM-1.pdf (Accessed on 19 April 2025).

C26. M. Zubair and Q. M. N. Amjad, "Uncertainty Reduction Using Multivariate Reliability Models," *2017 14th International Bhurban Conference on Applied Sciences and Technology (IBCAST)*, Islamabad, Pakistan, pp. 361–367, 2017.

C27. ISO/IEC/IEEE International Standard – *Systems and Software Engineering – Vocabulary*, ISO/IEC/IEEE 24765:2017, 2nd Ed., pp.1–418, 15 Dec. 2017. https://www.iso.org/standard/71952.html (Accessed on 24 July 2025).

C28. A. Loquercio, M. Segu and D. Scaramuzza, "A General Framework for Uncertainty Estimation in Deep Learning," *IEEE Robotics and Automation Letters*, vol. 5, no. 2, pp. 3153–3160, April 2020.

C29. V. Witkovský, G. Wimmer, Z. Ďurišová, S. Ďuriš and R. Palenčár, "Brief Overview of Methods for Measurement Uncertainty Analysis: GUM Uncertainty Framework, Monte Carlo Method, Characteristic Function Approach," *2017 11th International Conference on Measurement*, Smolenice, Slovakia, pp. 35–38, 2017.

C30. J. J. Thiagarajan, B. Venkatesh and D. Rajan, "Learn-By-Calibrating: Using Calibration as a Training Objective," *ICASSP 2020 – 2020 IEEE International Conference on Acoustics, Speech and Signal Processing (ICASSP)*, Barcelona, Spain, pp. 3632–3636, 2020.

C31. J. Ding, S. Ma, W. Yuan, J. Liu, C. Jia and X. Cui, "A New Method for Measuring Calibration Factors of Microwave Power Transfer Standard," *2021 IEEE MTT-S International Wireless Symposium (IWS)*, Nanjing, pp. 1–3, 2021.

C32. X. Jinzhe and X. Jiulong, "Calibration Interval Optimization and Calibration Conclusion Risk Analysis on Automatic Test System," *2017 13th IEEE International Conference on Electronic Measurement & Instruments (ICEMI)*, Yangzhou, China, pp. 236–241, 2017.

C33. D. Makwana1, P. Engineer, A. Dabhi and H. Chudasama, "Sampling Methods in Research: A Review," *International Journal of Trend in Scientific Research and Development (IJTSRD)*, vol. 7, no. 3, pp. 762–768, May-June 2023.

C34. R. Wen, J. Huang and Z. Zhao, "Multi-Agent Probabilistic Ensembles with Trajectory Sampling for Connected Autonomous Vehicles," *2023 IEEE Globecom Workshops (GC Wkshps)*, Kuala Lumpur, Malaysia, pp. 2025–2030, 2023.

C35. J. Cook, S. Ur Rehman and M. Arif Khan, "Lightweight Cryptanalysis of IoT Encryption Algorithms: Is Quota Sampling the Answer?" *IEEE Access*, vol. 12, pp. 147619–147639, 2024.

C36. Z. He, K. Yang, X. Cai and H. Sun, "Calibration Sample Number Determined by Theory of Sampling Provide Threshold for Multivariate Model Building," *2019 IEEE 2nd International Conference on Micro/Nano Sensors for AI, Healthcare, and Robotics (NSENS)*, Shenzhen, China, 2019.

C37. D. Tsai, S. Worrall, M. Shan, A. Lohr and E. Nebot, "Optimising the Selection of Samples for Robust LiDAR Camera Calibration," *2021 IEEE International Intelligent Transportation Systems Conference (ITSC)*, Indianapolis, IN, USA, pp. 2631–2638, 2021.

4 Calibration Steps

4.1 INTRODUCTION

Calibration is an inevitable part of measurements. Therefore, there are strict rules and regulations for successful calibrations. In this chapter, cost of calibration, functions of calibration laboratories, and accreditation organizations are discussed. Guidance for paper standards for reliable calibrations is detailed.

4.2 COST OF CALIBRATION

The cost of calibration depends on what is being calibrated and who is calibrating. In simple cases where one-off instruments are involved, the cost can be less than a few hundred dollars, but complex cases can cost tens of thousands of dollars. Cost also depends on many other factors, as shown in Figure 4.1. Calibrations can be conducted on the premises of the device use or by calibration laboratories or carried out by in-house personnel or outsourced to third parties.[D1]

The purpose of calibration is to ensure that the decisions made about the product/service do not result in false acceptance (consumer risk) or false rejection (producer risk)

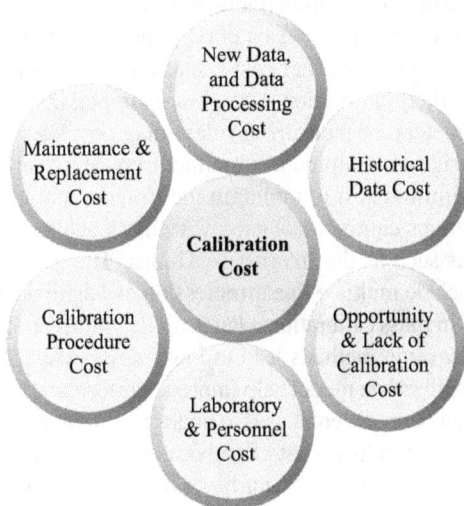

FIGURE 4.1 Cost of calibration. Calibrations can be very costly since it involves collection of data, processing of data, specialized laboratories, and well-trained personnel for successful calibrations. Some of the costs can be mitigated by regular calibration programs. Non-calibrated systems can impose further costs if they fail due to lack of calibration or in the case of local or international legal challenges.

DOI: 10.1201/9781003590767-4

of products or services, which leads to cost vs safety issues. In many situations, such as weighing systems, calibration is a statutory requirement.

Most calibration systems have a validity period during which the devices can be used free from concern of major errors and uncertainties. Some organizations carefully work out detailed methods for determining the calibration needs of their devices, while in others, calibration takes place barely to meet the legal requirements. In some organizations, the perception exists that cost is reduced if the calibration intervals can be stretched legitimately, whereas this approach can be very costly. Uncalibrated devices can have a negative effect on product quality, thus opening avenues for legal challenges as well as possible loss of market share to competitors. It is possible that the planned calibration costs can be, say, $200, as opposed to an unexpected failure of equipment, which may cost thousands or even millions of dollars.

Several different mathematical techniques, such as the Weibull statistical analysis and renewal equations, can be employed to analyze the costs. Weibull distribution is a probability distribution model using various distributions to interpret life data for identification of the key reliability matrix. Weibull analyses are used on collected data to optimize calibration intervals and determine associated costs. It predicts the reliability of equipment and helps to determine maintenance plans. There are many commercial software programs that can be selected (e.g., visualSMITH, Calibration Manager, and so on) for total cost analysis and determining the calibration costs.[D2] [D3]

A device must be calibrated if the failure rate increases or functionality noticeably deteriorates compared to other functional devices. As a rule of thumb, 85%–95% of all instruments returned for calibration must meet outgoing calibration limits determined by the probability chart of age and failure rates.

Calibration of instruments in manufacturing or service industries can be automated. In some cases, calibrations are batched to be carried out on multiple similar devices at once to save costs. In other cases, mass manufacturing of devices can be much less costly than their calibration costs. For example, instruments such as bathroom scales, thermometers, or measuring rulers may cost less than their calibration needs. Also, some original equipment manufacturers offer calibration services for their products with minimal cost to maintain their market competitiveness.

Lower-cost calibrations can be associated with mass calibration of devices relying on data-based systems such as the Internet of Things. The use of suitable algorithms and AI models that include multiple linear regression and artificial neural networks has proven to be effective in mass calibrations. Particularly, multi-parameter calibrations by supervised machine learning methods lead to lower costs. Multiple variables included in a multi-parameter calibration model help improve performance and produce accurate results. Integrating data from different sensor nodes into a calibration model using sensor fusion algorithms also results in cost-effective, consistent, accurate, and useful information in predicting the target variables far beyond calibration of individual sensors.[D4]

4.3 PROCEDURES AND DOCUMENTATION

The calibration process transfers a reference value to International System (SI) units to establish an "unbroken chain of comparisons" required for traceability. Due to sensitivity, the calibration procedure must be documented carefully, verified, and validated. The procedure is executed as a set of sequential operations in accordance with relevant fundamental and published standards. Therefore, the process involves many administrative and procedural issues, traceability, accreditation, quality assurance, and documentation.

Calibration documentation contains information on the accuracy and uncertainty of the measuring devices in addition to device models and serial numbers, calibration procedures, calibration dates, special conditions of use, tolerances, and calibration results.[D5] [D6]

Calibration labels are attached to calibrated devices. Labels need to conform to ISO/IEC 17025 General Requirements for the Competence of Testing and Calibration Laboratories.[A9] ISO 17025 requires that:

- All measurement equipment shall be securely and durably labeled,
- The labels should clearly indicate the name of the calibration laboratory, date of calibration, due date, usage equivalent, and the authorized personnel,
- Information on the label must be legible and durable under reasonable use and storage conditions,
- When it is impractical to affix a label directly on an item, the label may be affixed to the instrument container,
- Temper resistance labels may be used, when necessary, and
- Functional labels should contain reference standards.

As an example, labels on radioactivity measuring instruments contain information on calibration parameters, applicable dose levels, type of radioactivity detected, and radioactive source used as illustrated in Figure 4.2.[D7] [D8]

Devices with multiple probes have additional labels for each probe containing information such as:

- Date last calibrated and call date,
- Probe or sensor type,

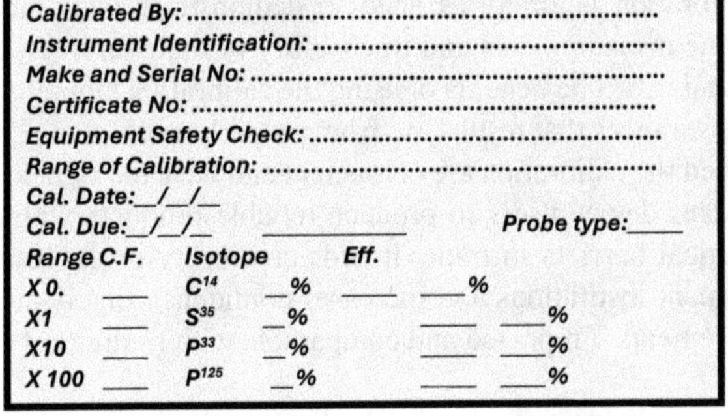

| Calibrated By: .... |
| Instrument Identification: .... |
| Make and Serial No: .... |
| Certificate No: .... |
| Equipment Safety Check: .... |
| Range of Calibration: .... |
| Cal. Date:__/__/__ |
| Cal. Due:__/__/__   Probe type:____ |
| Range C.F.   Isotope   Eff. |
| X 0.   ____   C¹⁴   __%   ____ ___% |
| X1   ____   S³⁵   __%   ___% |
| X10   ____   P³³   __%   ___% |
| X 100 ____   P¹²⁵ __%   ___% |

FIGURE 4.2 A typical calibration label. Calibration labels must be displayed on devices. They should have sufficient information on the date of calibration, and types of calibration. Labels need to be firmly affixed in an indestructible and temper proof manner.

- Information on organization and personnel who conducted the calibration,
- Range of measurement and correction factors for each range, and
- Efficiency of the instrument, etc.

Calibration labels are made from various materials such as metal plates, adhesive aluminum, vinyl overlays, or transparent adhesive tapes. In the case of metal plates and adhesive aluminum, information and markings may be permanently indented for durability. A calibration label may have the following additional information:

- Calibrate before use
- Not calibrated
- Obsolete
- Do not use
- Does not conform
- Indication only
- Information only
- Out of calibration
- Reference only
- Uncalibrated instrument
- User-calibrated instrument

4.4 ACCREDITATIONS AND ORGANIZATIONS

There are many accredited organizations to perform calibrations (Table 4.1). In the USA alone, the number of accredited organizations exceeds 1,640 operating in all the states.[A6] Accredited laboratories calibrate devices on the principles of traceability within the international reference standards ISO17025 requirements.

The accreditation organizations and the accreditation process are both independent and impartial. Accreditation requires operations within a legal framework and is completely unbound from commercial motivations. Accreditation is regarded as a public-authority activity. The accreditation process and organizations do not compete, nor affiliated to any third-party conformity-assessment bodies.[D9] [D10]

Test and calibration laboratories need to demonstrate their independence and impartiality in the measurements and traceability of their services, complying with international standards. The benefits of using the calibration laboratory by an organization include assurances that results are from properly calibrated equipment and the staff who executed the calibration are competent and have the right level of expertise. Calibration enables device users to produce reliable products with confidence and eliminates technical barriers in trade. It adds credibility to the results required by law and government institutions and increases confidence that data generated complies with requirements if repeated and comparable with findings of other accredited laboratories.

Some of the worldwide accreditation bodies are the International Accreditation Forum (IAF), the International Laboratory Accreditation Cooperation (ILAC), the European Cooperation for Accreditation, the Inter-American Accreditation Cooperation (IAAC), and the Asia Pacific Laboratory Accreditation Cooperation

TABLE 4.1

Partial List of Accredited Calibration Organizations in the USA and around the World

Accura Calibration	https://accuracalserv.com
Anritsu	Products & Solutions I Anritsu Asia Pacific
Calibration Laboratory, LLC	Calibration Laboratory, LLC
Certified Measurements, Inc	Certified Measurements, Inc.
Continental Testing	Continental Testing Services
Custom Calibration Inc	Onsite Calibration I Custom Calibration I United States, CT
Fukuda	Accredited Calibration Laboratory ISO/IEC 17025
Honeywell Process Solutions	About Honeywell Process Solutions
Intech Calibration	Home - Intech Calibration
KOMPASS India	Kompass
Mitutoyo	Mitutoyo Australia - Micrometers
Nagase & Co. Ltd.	NAGASE & CO., LTD.
National Instruments Corp.	Test and Measurement Systems, a part of Emerson - NI
Rohde & Schwarz	Rohde & Schwarz Australia I Rohde & Schwarz
SGS China	SGS China I We are the world's leading Testing, Inspection and Certification company.
Tektronix, Omega Eng., Inc	OMEGA ENGINEERING Calibration Services I Tektronix
Werner Bayer GmbH	Home I Machines I Processes and technologies I Leak test instruments I Process monitoring I Company I Contact I

(APLAC). A global acceptance of the services provided by these bodies is established by signing multilateral agreements, which construct essential trust mechanisms.

The safety and quality of trade is controlled by international standards and compliance programs that facilitate the movement of goods and services across international borders. For example, the European Commission has taken the path to prevent or restrict the marketing or use of products posing a serious risk to the health and safety of consumers. The only exception is for food, pharmaceutical, and medical devices since they are covered by other, much tighter mechanisms. Worldwide accreditation institutes are often reviewed. Conformity Europe (CE) marking, procedures, and standards for accreditation, certification, and calibration are summarized through tests and measurements

The International Organization for Standardization (ISO) is the world's largest standard-setting body. Many of the ISO's standards include guidance on the compliance of a service, a product, a person, or a system to meet the standards. ISO/IEC 17011 (Conformity Assessment—General Requirements for Accreditation Bodies Accrediting Conformity Assessment Bodies) defines accreditation as "a third-party attestation related to a conformity assessment body conveying formal demonstration of its competence to carry out specific conformity assessment tasks.[D11]

Accreditation in health, safety, and medical devices is continually gaining importance due to wider proliferation of medical and health equipment. The World Health Organization (WHO) considers medical calibration measurements as part of quality health care. WHO publishes guidelines for standard practice and seeks to ensure that healthcare providers focus on these issues. Properly calibrated equipment ensures accurate and reliable measurement data, which directly impacts the quality of care for patients. Each manufacturer is obligated to ensure inspection and testing their equipment to the highest level possible, including mechanical, automated, or electronic equipment. They should all be suitable for their intended purposes and can produce valid results.[D12]

Medical devices play a pivotal role in modern healthcare, aiding accurate diagnosis, monitoring, and treatment of patients. The reliability and accuracy of these devices are indispensable for ensuring patient safety and well-being. In the absence of proper calibration, there is a significant risk of incorrect readings, misdiagnoses, and inappropriate treatment plans, all of which can lead to harm. Regular calibration is essential not only for maintaining patient safety but also for meeting regulatory requirements and quality assurance standards. Compliance with calibration protocols ensures that medical institutions and professionals can deliver reliable and effective care. Also, calibrated medical devices enhance the efficiency and cost-effectiveness of healthcare systems by minimizing the need for repeated tests and procedures due to possible inaccurate results.[D13]

Medical devices may be classified as preventive care devices, assistive care devices, diagnostic devices, and therapeutic devices. Safety concerns and diversity of products manufactured in different countries, coupled with the sheer number of different devices in the market that demand the development of reliable devices. The WHO estimates that there are an estimated 2 million different kinds of medical devices on the world market that can be categorized into more than 7,000 generic device groups. Coming up with regulatory schemes is a serious challenge for domestic as well as international regulatory bodies.[D14] [D15]

4.5   CALIBRATION LABORATORIES

Calibration is conducted by organizations performing tests in permanent, temporary, or remote locations. Some organizations have several *calibration laboratories* to cater to different types of devices. Many organizations cover a broad spectrum of calibration technologies such as communications, defense, medical, aerospace, manufacturing, automotive, and chemical industries. Laboratories of these organizations are accredited by authorities in accordance to the guidelines set out, such as the ISO/IEC 17025:2018. Accreditation is a formal recognition that a particular laboratory is competent to conduct specific tests and calibrations.[D16]

Laboratory activities are undertaken impartially in their structures and management styles. The laboratory is responsible, through legally enforceable commitments, for the correctness and validity of all information obtained or created during the performance of its activities.

A laboratory may offer calibration services under different categories such as a *permanent laboratory*, *onsite facility laboratory*, and *mobile facility laboratory*. Irrespective of the types of services selected, a laboratory should have the right personnel, facilities, equipment, procedures, and support services necessary to manage and perform its activities. Laboratories establish and maintain traceability of their measurement results by means of a documented unbroken chain of calibration processes, each contributing to the measurement uncertainty, linking them to appropriate references. Laboratories ensure that measurement results are traceable to the International System of Units (SI) through (1) calibrations are provided competently, and (2) certified reference materials are provided by a competent producer with a proven traceability to the SI.

The laboratory may have a preset procedure for transportation, receiving, handling, protection, storage, retention, disposal, or return of calibration items, including all provisions necessary to protect the integrity of the calibrated and to protect the interests of the laboratory and the interests of the customers.

Calibration laboratories maintain quality manuals for defining and identifying policies, procedures, requirements for competence, and consistency in operations that comply with the requirements of ISO/IEC 17025.

In an organization, establishing and maintaining laboratory quality control is essential. Quality control mechanisms are given as follows:

- *Internal quality control (IQC)*: Laboratory performs IQC checks for all relevant tests and procedures.
- *External quality assessment (EQA)*: Laboratories participate in interlaboratory comparisons in national, regional, and international EQA schemes.
- *Internal audits*: This is the process of critical review of laboratory activities by managers or delegated persons or teams.
- *External audits*: Customers may request an external person, usually a laboratory technologist or laboratory scientist, to review the laboratory quality management (QMS).
- *Management review*: At least once a year, the head of the laboratory and the QM review the services provided by the laboratory and the established QMS.
- *Continuous quality improvement (CQI)*: All operational procedures must be systematically and continuously reviewed by the head of the laboratory to identify potential sources of noncompliance and the areas that require improvement.

4.6 PERSONNEL

The calibration is conducted by trained personnel using appropriate devices to implement the reference standards. The devices are kept as secondary standards and working standards by national authorities and companies, respectively. The calibration process involves a set of rules, regulations, and processes. During calibration, the calibrating personnel observe the following:

1. The preparation procedure is consistent.
2. Test items and reference standard respond in the same manner in the test environment.
3. Random errors associated with the measurement are independent.

4. Multiple measurements form a distribution curve with the same standard deviation.
5. The test item and reference standards are stable during the measurement.
6. The bias of the test item and reference standard are the same, and hence it can be eliminated by taking further readings.
7. Once the calibrations are carried out, ideally, the difference should be zero or the difference should be explainable by appropriate charts, curves, and statistical relationships.
8. Calibration is repeatable, yielding the same results.

Calibration personnel should have appropriate clothing (e.g., static-free clothing, gloves, face masks, radiation protective, etc.) for safety as well as for avoiding the possibility of contamination of the calibrated devices. Although fully trained and experienced, it is possible for different operators to produce measurements differing from the previous ones in terms of signs and magnitudes. To overcome this problem, measurements by different operators can be assessed, plotted, and compared. Alternatively, calibration is repeated by the same operator, and calibration curves and charts are compared and analyzed independently. This may not be a problem with automated calibrations.

Calibration personnel are usually required to have a bachelor's degree in related fields such as electrical, electronics, biomedical engineering, instrumentation engineering, or postgraduate degrees in relevant fields. The personnel should have a good knowledge of basic principles of testing, evaluation of calibration results, use of software, equipment capabilities, and uncertainty of test results, applicable regulations, and general requirements expressed in relevant and associated standards.

Laboratories keep copies of the instruction manual and a service manual with the equipment. The service manual should be available for each equipment, to provide the technical information needed by the engineers in charge. Service manuals are mandatory for technical specifications during the procurement of equipment. The IEC 60601-11, for example, determines that all equipment must be accompanied by a description of calibration techniques supplied by the manufacturer.[D17] [D18]

The purpose of keeping documents properly is to meet requirements within which a laboratory operates to demonstrate its competency to carry out calibration, complying with international standards such ISO/IEC 17025, ISO 13485, IEC 60601, IEC 62353, and Global Harmonization Task Force. Qualified personnel with experience are engaged in selection and maintenance of calibration equipment.

4.7   COMPARISON OF PRE- AND POST-ADJUSTMENTS

The performance of a device with an unknown accuracy is initially compared with the performance of other device(s) with known accuracy. The deviation of results of accuracy indicates the necessity for calibration or replacement of that device. Comparison is also conducted after the calibration to ensure the expected level of accuracy is achieved. In some cases, pre-adjustments are carried out to prepare the equipment under calibration for the environmental conditions they are operating. Post-adjustments are also necessary to account for transportation and packaging conditions if external calibration services are involved. Pre-adjustments and post-adjustments are backed up with appropriate software tools and analytical algorithms.[D19] [D20] [D21]

Normally, measuring systems are calibrated against primary standards via secondary or working standards. In some cases, comparison between measurements of two devices can be viewed as a form of calibration if one of them has already been calibrated and demonstrates a reliable level of accuracy and certainty.

4.8 PREPARATION FOR CALIBRATION

Preparation for calibration is essential since it requires the determination of critical errors and uncertainties associated with the measurements. The deviation of a device reading from the conventional "true value" of the measurand is established and documented by identifying all possible affecting conditions. A good example of preparation for calibration is the measurement assurance planning.

Measurement assurance programs are groups of activities designed to establish frameworks to ensure reliability and accuracy of measurements. Measurement programs are applied by the National Aeronautics and Space Administration (NASA) and National Institute of Standards and Technology (NIST) for describing to calibration of instruments for their Earth Observing System (EOS) operations. These programs are applied for the calibration of the EOS satellite and ground-based and airborne instruments. They consist of a series of carefully designed activities in space exploration and measurements.[D22]

Measurement assurance programs are applied to achieve accurate, precise, and consistent long-term reliable measurement data from multiple instruments and multiple platforms. Many large and small projects depend on the proper calibration of all instruments against a set of recognized physical standards and carefully characterizing the instrument performance at the system level. Other important points would be adhering to good measurement practices and established protocols, intercomparing measurements wherever and whenever possible. Establishing traceability for all instruments to a common scale of physical quantities is essential.[D23]

4.9 CALIBRATION STANDARDS

Standards ensure desirable characteristics of products and services such as quality, safety, reliability, efficiency, and reproducibility. ISO standards provide a technical base for health, safety, and conformity assessment of devices. The ISO standards are broadly accepted in the world. However, the acceptance is voluntary since ISO has no legal authority to enforce its standards.

There are numerous international standards for calibration-related practices; some of these are listed below:

- ISO/IEC/EN 17025 is the general requirements for the competence of testing and calibration laboratories.
- BS/EN/ISO 9000:2000 is a family of standards for international quality management. These standards are globally accepted and applied by many organizations.

- ANSI/NCSL (Z540) standard sets compliance guidelines for calibration laboratories and measuring and test equipment.
- ISO 9001 standard sets the quality management system requirements.
- MIL-STD-45662A is a military grade standard that describes the requirements for creating and maintaining calibration systems for measurement and testing.
- MSL 904001 outlines the ability to calibrate and test equipment in accordance with standard calibration procedures and documentation.
- ISO/IEC 17034 Conformity assessment sets the general requirements for proficiency testing.
- ISO 9846 and ISO 9847 standards are set for the calibration of radiometers for the measurement of broadband solar radiation.
- ISM standards are a set of standards aiming industries for testing.
- NIST standards.
- ISO9100 standardized quality management in the aerospace industry.
- *IATF 16949:2016* is a quality management standard in the automotive industry.
- ISO 15189/ISO 13485 Medical laboratories/Medical Devices outlines particular requirements for quality and competence.
- ISO 13528 outlines statistical methods for use in proficiency testing by interlaboratory comparison.
- ISO 13485 is for calibration requirements/data loggers for compliance.
- OECD GLP outlines OECD principles on good laboratory practice.
- ISO Guide 34 outlines the general requirements for the competence of reference material producers.
- ISO 8402 sets the quality management and quality assurance—vocabulary.·
- ISO 19011 sets the guidelines for quality and/or environmental management system auditing and many others.

ISO/IEC 17025 standards play an important role in supporting the validity and reliability of results generated by calibrations. ISO/IEC 17025 facilitates international cooperation by establishing wider acceptance of test results between countries and other organizations. This not only provides global confidence in the work of an organization but also promotes national and international trade through the acceptance of calibrations between countries without the need for additional testing.

The internationally accepted standards of the ISO 15189 and ISO 17025 have stringent requirements and can usually be met only by laboratories either at the national level or by specialized reference laboratory level. They are often very resource-intensive, and many countries find it difficult to establish and maintain such laboratories.[A7] In health-related device calibrations are much more stringent quiring strict administrative procedures outlined by ISO 9000 series of standards. ISO 15189 was officially launched in 2004, and it includes both management as well as technical requirements for medical laboratories and laboratory procedures.

4.10 CONCLUSIONS

This chapter discusses calibration steps in detail. It has been shown that calibration can be a very costly exercise and cannot be avoided. The sheer number of calibration accreditation organizations, existence of many laboratories, and diverse range of standards indicate the seriousness of calibrations. A partial list of paper standards related to calibrations is provided.

REFERENCES

D1. S. Andonov and M. Cundeva-Blajer, "Comparative Cost and Benefit Analysis of TCal and Classical Calibration," *2020 IEEE International Workshop on Metrology for Industry 4.0 & IoT*, Roma, Italy, pp. 69–74, 2020.

D2. J. Zhao, G. Peng and H. Zhang, "Schedule and Cost Integrated Estimation for Complex Product Modeling Based on Weibull Distribution," *2015 IEEE 19th International Conference on Computer Supported Cooperative Work in Design (CSCWD)*, Calabria, Italy, pp. 276–280, 2015.

D3. N. Li, Q. Liu, L. Wang and X. Liu, "A Novel Approach for Automation of Precision Calibration Process," *2008 4th International Conference on Information and Automation for Sustainability*, Colombo, Sri Lanka, pp. 319–323, 2008.

D4. N. U. Okafor and D. T. Delaney, "Application of Machine Learning Techniques for the Calibration of Low-cost IoT Sensors in Environmental Monitoring Networks," *2020 IEEE 6th World Forum on Internet of Things (WF-IoT)*, New Orleans, LA, USA, pp. 1–3, 2020.

D5. J. Mitchem, "Automated Calibration Aids Smooth Turnover of New Plants," *IEEE Instrumentation & Measurement Magazine*, vol. 7, no. 4, pp. 26–29, Dec. 2004.

D6. W. Wang and C. Cao, "NOAA-20 VIIRS Sensor Data Records Geometric and Radiometric Calibration Performance One Year in-Orbit," *IGARSS 2019 – 2019 IEEE International Geoscience and Remote Sensing Symposium*, Yokohama, Japan, pp. 8485–8488, 2019.

D7. Y. C. Lim and M. Kang, "Target Round Bar Detection and Two-Stage Calibration for Automatic Label Attachment System," *2023 5th International Conference on Control and Robotics (ICCR)*, Tokyo, Japan, pp. 225–229, 2023.

D8. C. Brown et al., "Infrastructure for Digital Calibration Certificates," *2020 IEEE International Workshop on Metrology for Industry 4.0 & IoT*, Roma, Italy, pp. 485–489,2020.

D9. L. Sevgi, "Accreditation: Crucial in World Trade, Public Safety, and Human Rights," *IEEE Antennas and Propagation Magazine*, vol. 56, no. 4, pp. 265–275, Aug. 2014.

D10. R. Benitez, C. Ramirez and J. A. Vazquez, "Sensor Calibration for Metrology 4.0," *2019 II Workshop on Metrology for Industry 4.0 and IoT (MetroInd4.0&IoT)*, Naples, Italy, pp. 296–299, 2019.

D11. ISO/IEC 17025 – General requirements for the competence of testing and calibration laboratories. https://www.iso.org/standard/66912.html (Accessed on 19 April 2025).

D12. WHO, *Anex 3*, Good Manufacturing Practices: Guidelines on Validation, trs1019-annex3-gmp-validation.pdf (Accessed on 19 April 2025).

D13. R. Kumar, R. K. Choudhary, Archana, and N. K. Yadav, "Calibration of Medical Devices: Method and Impact on Operation Quality," *International Journal of Pharmaceutical Sciences*, vol. 16, no. 1, pp. 1–13, Sept. 2023.

D14. M. Sezdi, "Expansion of Authorization Groups in Testing, Control and Calibration Activities of Medical Devices," *2022 Medical Technologies Congress (TIPTEKNO)*, Antalya, Turkey, pp. 1–4, 2022.

D15. B. Karaböce, "Challenges for Medical Metrology," *IEEE Instrumentation & Measurement Magazine*, vol. 23, no. 4, pp. 48–55, June 2020.

D16. ISO/IEC 17025:2018, Laboratory Quality Manual. https://abclabs.com.au/assets/laboratories/1.1.%20Laboratory%20Quality%20Manual.pdf (Accessed on 19 April 2015).

D17. IEC 60601-1-11:2015, Part 1–11: General Requirements for Basic Safety and Essential Performance — Collateral Standard: Requirements for Medical Electrical Equipment and Medical Electrical Systems Used in the Home Healthcare Environment https://www.iso.org/standard/65529.html (Accessed on 19 April 2025).

D18. E. Wu, K. Wu, R. Daneshjou, D. Ouyang, D. E. Ho and J. Zou, "How Medical AI Devices Are Evaluated: Limitations and Recommendations from an Analysis of FDA Approvals," *Nature Medicine*, vol. 27, no. 4, pp. 582–584, April 2021.

D19. A. J. Rousseau, T. Becker, J. Bertels, M. B. Blaschko and D. Valkenborg, "Post Training Uncertainty Calibration of Deep Networks for Medical Image Segmentation," *2021 IEEE 18th International Symposium on Biomedical Imaging (ISBI)*, Nice, France, pp. 1052–1056, 2021.

D20. K. Imaoka, Y. Fujimoto, M. Kachi, T. Takeshima, K. Shiomi, H. Mikai, T. Mutoh, M. Yoshikawa and A. Shibata, "Post-Launch Calibration and Data Evaluation of AMSR-E," *IGARSS 2003. 2003 IEEE International Geoscience and Remote Sensing Symposium. Proceedings (IEEE Cat. No.03CH37477)*, Toulouse, France, vol. 1, pp. 666–668, 2003.

D21. M. W. Spencer, W.-Y. Tsai and G. Neumann, "NASA Scatterometer (NSCAT) Pre-Launch Calibration Results and Post-Launch Calibration Plan," *1995 International Geoscience and Remote Sensing Symposium, IGARSS '95. Quantitative Remote Sensing for Science and Applications*, Firenze, Italy, vol. 1, pp. 821–823, 1995.

D22. J. L. Daniels, G. L. Smith, K. J. Priestley and S. Thomas, "Using Lunar Observations to Validate In-Flight Calibrations of Clouds and the Earth's Radiant Energy System Instruments," *IEEE Transactions on Geoscience and Remote Sensing*, vol. 53, no. 9, pp. 5110–5116, Sept. 2015.

D23. C. Serief, N. Khorchef and Y. Ghelamallah, "Modeling of Temperature Rises at Focal-Plane-Array and Their Impact on the Performance of a CCD-Based Spaceborne Earth-Observing Imaging System," *IEEE Transactions on Device and Materials Reliability*, vol. 24, no. 2, pp. 335–343, June 2024.

5 Artificial Intelligence in Calibrations

5.1 INTRODUCTION

In this chapter, artificial intelligence (AI) models that have been applied in calibrations are discussed. This is an evolving area but finds diverse and novel applications. AI models attract much attention as they are expected to lead a new industrial revolution. AI relies on large data, and there is no shortage of data in today's technology. AI-related data-driven calibration methods have been applied in physics-based systems, in geology, climatology, medical devices, biology, healthcare, finance, cognitive science, science and engineering, and applied physics.

The terms machine learning (ML) and AI are used in this book in an interchangeable way. Nevertheless, the way these terms are used in the literature needs to be clarified at the beginning of this chapter. As the name indicates, ML is the learning of machines from data that focuses on learning patterns to make predictions and draw conclusions, whereas AI is ML too or learning of machines, but it has a broader coverage. AI performs tasks close to human intelligence and makes decisions closer to human decision-making. ML is usually configured by mathematical algorithms, which are a form of computing written down in program codes, whereas AI-related rules and behavior are inferred from training data, rather than written down in program codes. Therefore, AI is a much broader concept since it can include cyber-physical systems that autonomously learn, evolve, and adapt to new conditions over time, just like human beings. Some examples are robots, drones, autonomous vehicles, Internet of Things (IoT), and so on. In this book, the term AI is used more often than ML.

AI models can be purely based on software operating in a virtual world. Some examples are voice recognition, voice assistant search engines, image analysis, speech recognition and imitation, and face recognition are a few to mention. Or AI can be a combination of hardware and software embedded into devices such as robots, drones, and autonomous vehicles. AI models learn about their environment and take actions in an intelligent manner. AI is a data-driven system that requires large and changing datasets. It relies on robust and evolutionary infrastructures, firm ethics, and trustworthiness.

In calibration systems, AI methods are used to improve operational characteristics of individual or multiple devices in complex systems. Examples are weather forecasting in meteorology, control systems in plants, IoTs, medical applications, and many other large data-driven measurement systems. Data-driven calibration for accurate operations relies on the availability of large data, datasets, and ground truth data. The sampling techniques and data required from the devices for accurate calibrations increase proportionally with the increase in the complexity of a system.

DOI: 10.1201/9781003590767-5

5.2 HANDLING DATA IN DATA-DRIVEN CALIBRATION

AI models help to enhance mathematical modeling or eliminate the need for modeling, mathematical descriptions, and simulations of real-time operational devices. For example, supervised learning, which is a neural network algorithm, is a commonly applied method trained to determine relations between observed values and model parameters on a real-time basis. The accuracy of the methods is strongly dependent on the representative quality of the training datasets.[E1]

Successful calibrations depend on the proper generation and handling of data, as well as on the selection of appropriate sampling techniques shown in Figure 5.1. Calibration using AI requires large amounts of data, which may be available from the following sources:

1. Data generated by measuring devices (e.g., sensors and instruments)
2. Data stored in historical files and archives
3. Assimilated data
4. Data historians of processes, data lakes, and data silos
5. Data stored on the internet and cloud
6. Ground truth data from cyber-physical systems
7. Simulation data

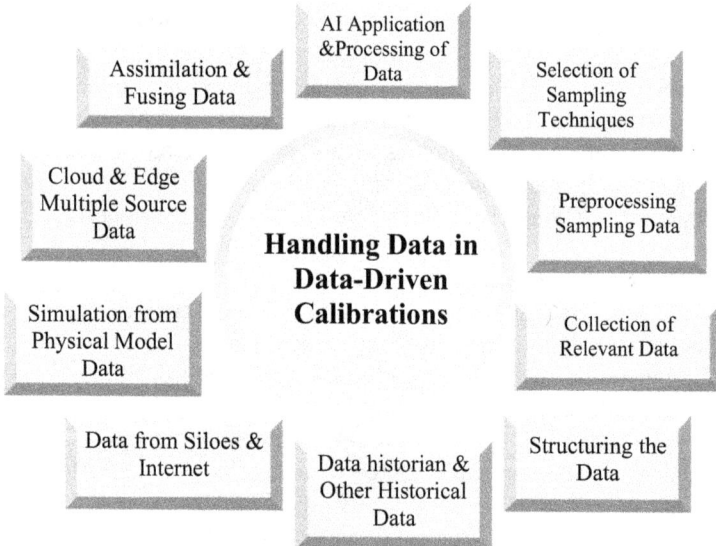

FIGURE 5.1 Factors in handling data in data-driven calibration. AI requires large data for training and inference. Data that can be used for calibrations comes from different sources. Apart from the data generated by devices under test, there are other sources such as data historians, data silos, cloud data, physical model data, ground truth data, simulation data, and so on. Collected data needs to be preprocessed to make it suitable for calibration.

Data generated by measuring devices: Many devices such as sensors and instruments generate large volumes of data in monitoring and measuring applications. Some operations collect several terabytes of data per day, as in the case of healthcare systems, weather forecasting, communication systems, space exploration, and research laboratories. A diverse range of burden data may be collected in addition to the useful data. This can present serious challenges in data collection, data handling, and processing of suitable data for calibration purposes. However, data-driven algorithms can help achieve high-quality, clean, versatile, and applicable data ready for calibrations.

Many operations have a broad range of measuring devices gathering data from multiple sources with different physical principles. Data from multiple sources can be combined into a single data frame to obtain a comprehensive dataset. Data from different sensors are integrated together to train an AI model for data transformation. *Data transformation* is the conversion of data into a suitable format for analysis and decision-making. It may include normalizing, balancing, scaling, and converting data from one type to another. Imbalanced datasets can lead to biased models during training, which can result in poor performance predictions, especially in under-sampling, over-sampling cases.[E2]

Data stored in historical files and archives: In many applications, a large amount of structured or unstructured data exist in datasets. Some of the data might have accumulated over decades. Still, these datasets find various AI applications for informed decision-making, as in the case of climate change studies. Accumulated data can contain missing values and incomplete data due to faults, failures, drifts, tampering, security attacks, connection errors, and other operational issues. When used for calibration purposes, data in these datasets should be handled for validity and completeness, as invalid data can have a substantial effect on calibration accuracy.[E3]

Many organizations practice file storage. Unless handled carefully, file storage methods may have large data redundancy and may contain substantial invalid data. The amount of calibration data increases over time due to the initial calibration and periodic calibrations. Calibration data accumulated over time may differ due to different times of calibration, variations in operational and device parameters, and conditions that existed during calibrations.[E4]

Assimilated data: Data assimilation is used when large volumes of useful data are needed. AI-based assimilation performs statistical assessments as well as data fusion. There are many off-the-shelf or custom-designed algorithms for a particular application that are capable of large-scale data assimilation. For example, in weather forecasting, the Global Data Assimilation System (GDAS) uses several thousand processors to perform optimization steps to match current forecasting analysis with the real-time observations. Another example is the Multi-Instrument Inversion and Data Assimilation Preprocessing System (MIIDAPS). AI version (MIIDAPS-AI) is a deep fully connected NN that describes a nonlinear mapping between measurements.[E5]

Data historians, data lakes, and data silos: Data historians, particularly in industrial applications, constitute large volumes of labeled data. Data historians are used to assess the health of sensors and instruments for validation and reconciliation of data for identifying incomplete or irregular measurements. Many organizations keep large amounts of data, generated in-house operations, termed as *data lakes* and *data silos*.[E6]

Data stored on the internet and cloud: The use of internet- and cloud-based data is well established, as in the case of a large language model (LLM) (ChatGPT, and many others). Some organizations allow for the use of calibration data available on the internet. For example, Calibration Executive by National Instruments can be used in either automatic or manual form. Various other databases are developed by NIST. For instance, the Information System to Support Calibrations (ISSC) is suitable for tracking and maintaining calibration information from processes and for generating calibration reports. Another example is the Calibration Check Standard Database (CCSD), which can be used for data storage, backup, and electronic retrieval of metrology-related data.[E7] [E8]

Cyber-physical ground truth data: Many physical processes are integrated on a real-time basis with computational algorithms. These operations are termed as cyber-physical systems. Cyber-physical systems provide seamless interaction between machines and humans. Typical examples of cyber-physical systems are autonomous vehicles, robotics, industrial automation, healthcare systems, IoT, and others. These processes generate information by direct observations and measurement, termed as the *ground truth data*. A large amount of ground truth data can be processed by AI models for calibration purposes.[E9]

Simulation data and data from datasets: Simulation data are synthetic data used to describe the real-world data as closely as possible. They can be generated from artificial datasets or from statistical or computational descriptions of real-world physical processes. Simulation data are used in statistics, computer science, and AI models to test algorithms, assess models, and conduct experiments. Simulation data find extensive applications in AI-related calibrations.

Data-driven techniques have been deployed primarily in non-parametric calibration applications where data are ubiquitous. AI helps to understand and model the real world by extracting patterns from a large dataset. Data-driven methods offer advantages in regressions, predictions, and autonomous decision-making.[B45] AI can be applied in supervised and unsupervised datasets to improve data quality. Methods like back propagation neural network (BPNN), extreme learning machine, and radial basis function neural network have been applied in numerous data-driven calibrations, yielding in good results. All data-driven techniques in calibration can be tested by the coefficient of determination (R^2), root mean square error (RMSE), mean absolute error (MAE), and other methods.[E10]

AI provides universal models in a wide range of applications based on a set of hardware. In many cases, it is convenient to compensate by hardware adjustments instead of compensating for errors by using software. However, data-driven models can minimize or eliminate adjustments and replacement needs of hardware components.

AI models can reformulate the characteristics of descriptive functions automatically under changing system conditions. This avoids the reformulation of specific mathematical descriptions. Similarly, artificial neural networks (ANNs) offer universal, accurate, and practical calibration information without the need for mathematical descriptions.

A data-driven calibration may use various methods such as Ridge regression, support vector regression, Gaussian regression, and others.[B46] [E11]

The use of AI models is revolutionizing the processing of large volumes of multidimensional datasets to extract useful information from. For example, onboard satellite sensors used in global observation systems generate large amounts of data regarding the temperature, humidity, pressure, cloud formation, and air movement in the atmosphere. Gathered information combined with land-based data on Earth's surface from oceans, lakes, and land is processed much more conveniently using AI. Despite the conveniences, training large-scale AI models requires elaborate computational resources and complex engineering skills to handle the data throughput and involves a diverse range of multimodal devices.[E12]

5.3   AI REGRESSION

Regression establishes the relationship among variables. It is an estimation technique for predicting and forecasting how variables affect each other. Predictions are evaluated by metrics such as variance, bias, accuracy, and error. *Variance* indicates the amount of changes in the estimated target function when different training datasets are used. In a good estimation, the target function should remain stable with very little variance. *Bias* is the tendency of the algorithm to learn the wrong things by not taking all the information into account. Inconsistency in the dataset such as missing values, errors in the input data, and insufficient data causes biases. *Error* is the difference between the actual value and the predicted value estimated by the model. *Accuracy* is the fraction of predictions that the model has got predictions right. An ideal model is expected to have a low variance, low bias, and low error.[E13]

Datasets are partitioned into training data and test data. Models learn patterns from the training data, and their performance is evaluated on the test data. The aim of training the models is to identify the patterns in the dataset, not necessarily mimicking or memorizing the training data. In training and testing, underfitting and overfitting may take place. *Underfitting* occurs if the model performs low accuracy on the test data and low accuracy on the training data. This is an indication that the model cannot capture the underlying patterns in the data to make meaningful predictions. *Overfitting*, on the other hand, is an indication of the model performing well on the trained data with low errors. The model learns the training data too well, including outliers and noise. This results in significantly high errors in predictions on the new datasets (Figure 5.2).[E14]

There are many regression methods used in calibration applications. Some of these methods are ANN, elastic net, linear regression, logistic regression, polynomial regression, random forest (RF) regressor, ridge regression, robust regression, stochastic gradient ascent/descent regression, SVM, and others. Most used AI models in regression analysis are Bayesian linear regression, decision tree regression, Lasso regression, negative binomial regression, ordinal regression, Poisson regression, principle component regression (PCR), quantile regression, RF regression (RFR), regularization, ridge regression, stepwise regression, SVR regression, Tobit regression, and others.[E15]

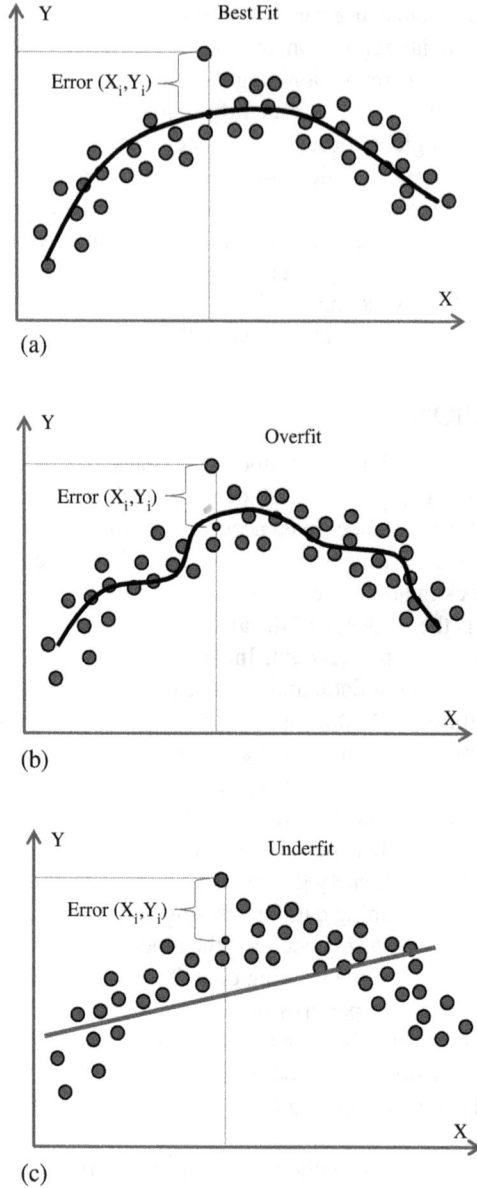

FIGURE 5.2   AI regression fits. AI is inherently suitable for regression and regression analysis. Depending on the selected model, training, and testing, regression fits could be (a) best fit, (b) overfit, or (c) underfit regression. A good fit indicates that the model selected learns the patterns rather than memorize them.

Linear regression is discussed in Chapter 3. Here, we will give logistic regression as an example. Logistic regression is a method for solving categorical classification problems that uses linear regression inside a sigmoid or logit function

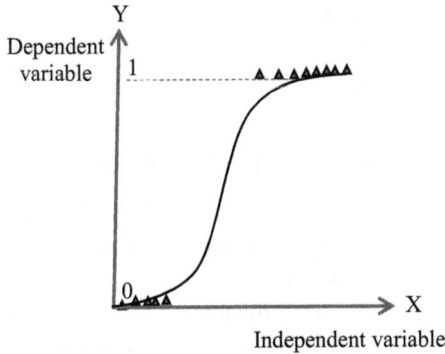

FIGURE 5.3 Logistic regression as a sigmoid function. Logistic regression predicts the probability of binary (crisp) outcomes as one or zero. It is used for estimating the probability of a particular outcome occurring by indicating the strength and direction of the relationship between independent variables.

(see Figure 5.3). It compresses the values to a range of 0 or 1 (Sigmoid function), resulting in a crisp probability. Logistic regression is often used as the first method in categorical predictions.[E16] [E17]

The equation for logistic regression is the sigmoid function (activation function) that can be expressed as:

$$f(x) = \frac{1}{1+e^{-x}} \qquad (5.3.1)$$

Where e = natural logarithm base, and x = numerical value to be transformed.

Multivariate analysis requires somewhat tailored methods such as linear regression, PCR, partial least squares (PLS), multi-layer perceptron (MLP), convolutional neural network (CNN), RFR, and others. Two typical examples are principal component analysis (PCA) and PLS. PCA finds the regression coefficients for a set of variables, and PLS establishes a linear regression model by protruding x and y variables to a new space. This technique is mainly used when the number of data points is less than the number of variables.[E18]

Once the regression is completed, further analysis may be necessary for validation of the results. Validations can be done by *graphical inspections,* which are recommended by most of the validation guidelines. Other methods of validation are as follows: *nongraphical approach* by visual inspection of data without plotting the graph or using statistical tools. *Statistical assessment,* which is the statistical evaluation of data. *Analysis of variance (ANOVA)* calculates combined variances. *Lack-of-Fit (LOF) test* is a statistical model to determine the accuracy of the relationship between variables. *Mandel's fitting test* is used to compare two models. In addition, *numerical assessment* and *numerical fitting parameters* are used as a measure of goodness-of-fit (GOF). The *correlation coefficient (r) and coefficient of determination (R^2)* are used to express the GOF of a model. *Residual standard deviation* indicates that the smaller the value is, the better the fit.[E19]

Typical other metrics often used for comparison of AI model performances are MAE, accuracy (%), R^2, Pearson correlation coefficient, Kullback–Liebler divergence, Jenson–Shannon divergence, and so on.

5.4 AI CLASSIFICATION AND CLUSTERING

Classification is one of the crucial components of AI. Classification is the organization of data into predefined groups or classes. It enables efficient data retrieval and analysis by assigning class labels to input data. Types of classification tasks are binary classification, multi-class classification, multi-label classification, and imbalanced classification. There are numerous classification algorithms. Some of these algorithms are decision trees, Kernel approximation algorithm, logistic regression, Naive Bayes (NB), K-nearest neighbor (KNN), RF, stochastic gradient descent (SGD), support vector machine (SVM), and others. As an example, a naïve bias classification is illustrated in Figure 5.4.[E20]

NB is based on conditional probability. Gaussian, multimodal, and Bernoulli models are the three major models used. NB uses a probability table based on feature values. The fundamental assumption is a conditional dependence, hence the term Naïve. It assumes that all features of the input are dependent on each other. The naïve Bayesian algorithm makes large datasets easier to handle, and it is an effective method for classification. The Bayes theorem provides a basis for later probability calculations. A simple formula used in NB is as follows:

$$P\left(\frac{A}{B}\right) = \frac{P\left(\frac{B}{A}\right)P(A)}{P(B)} \tag{5.4.1}$$

Where
  A, B = events
  $P(A/B)$ probability of A given B is true,
  $P(B/A)$ probability of B given A is true, and
  $P(A)$ and $P(B)$ are independent probabilities of A and B.

Clustering is a way of simplifying data by grouping the data based on similarities. Clustering identifies the groups of similar objects in datasets, thus revealing hidden structures to assist decision-making. Basic types of data clustering are density-based clustering, centroid-based clustering, distribution-based clustering, and hierarchy-based clustering. Numerous algorithms have been developed based on these basic types. These are agglomerative balance iterative reduction clustering, centroid-based clustering, constraint-based clustering, density-based clustering, distribution-based clustering, expectation minimization, fuzzy clustering, hierarchical-based clustering, fuzzy C-means clustering, Gaussian mixture model, K-means clustering (KMC), partition-based clustering, subtractive clustering, and others.[E21]

An example is the K-means clustering algorithm. KMC classifies an object into K-number of objects based on the features and attributes. Grouping can be carried out by reducing the number of squares of distances between data and the corresponding centroid, as shown in Figure 5.4. This technique uses the Euclidean distance

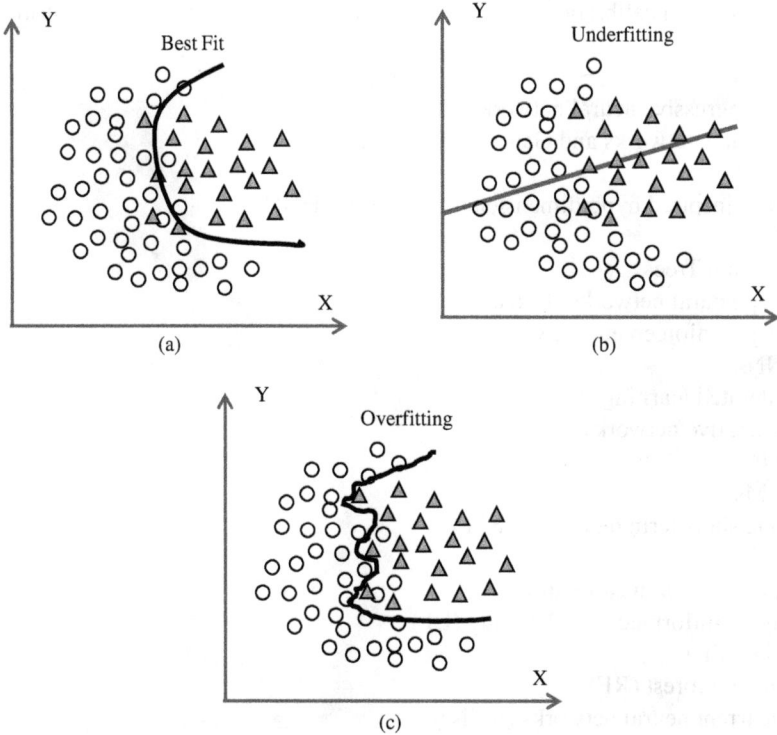

FIGURE 5.4 AI clustering. AI clustering is a process of organizing data into subgroups with similar elements or attributes. Clustering helps calibrations by sorting out similar data coming from different sources or similar data generated by the same device, in which other forms of data exist, such as noise. Clustering finds applications in large data and datasets that contain information other than information needed for calibrations. Depending on the model used, clustering can exhibit (a) best-fit cluster, (b) underfit cluster, or (c) overfit cluster.

metric to divide n observations into k clusters. Its main objective is to minimize the variance between the data and the corresponding centroid of the cluster. K-means clustering is useful for feature learning and for providing a starting point for other algorithms.[E22]

5.5   AI IN CALIBRATIONS

AI is a growing area in calibration and management of data generated from measuring devices. An example of AI applications is the SVM regression, which is used for calibration and improving the performance of large sensor systems. SVM can also be used for regression and distribution estimation. Other examples are linear regression models, feedforward neural network (FNN). Linear regression is a simple but important method to assess the linearity of relationships in predictions and in analysis. AI techniques are applied in linear and nonlinear calibrations for supervised and unsupervised learning.[E23] [E24]

AI models in calibrations can be listed as below, but this list is not exhaustive:

ANNs
Autoregressive neural networks
Bayesian networks and Gaussian naïve Bayes
BPNNs
Brain-inspired hyperdimensional computing (HDC)
CNN
Decision Trees
Deep neural networks (DNNs)
Deep reinforcement networks
FNNs
Federated learning (FL)
Generative networks
KNN
LLMs
Long short-term memory (LSTM)
MLP
Optimization in calibration
Physics-informed deep learning (DL)
Q-learning
Random forest (RF)
Recurrent neural networks (RNNs)
Semi-supervised networks
Statistical methods and AI
SVM

Most relevant methods with calibration studies and applications will be explained next.

5.6   ANN IN CALIBRATIONS

ANN is an interconnected group of nodes known as artificial neurons arranged in layers. Neurons are the processing elements that connect input and output layers. Strengths of connection are determined by the adjustable weights in the neurons. Neurons are given initial weights to be adjusted and readjusted by the training data supplied from the input layer. Once the weights in one layer reach a specific threshold level, then its outputs are transmitted to the next layer that holds the next set of nodes, as shown in Figure 5.5.

ANNs are trained by learning algorithms. Algorithms use *local search* methods (such as Hill climbing, local beam search, simulated annealing, and so on) to choose the weights that will result in the right outputs for the training inputs. Weights are adjusted to meet the preset estimation requirement within some acceptable level of errors.[E25]

The neural network is trained by calculating the difference between the actual output and the desired output. For this, optimization techniques are used.

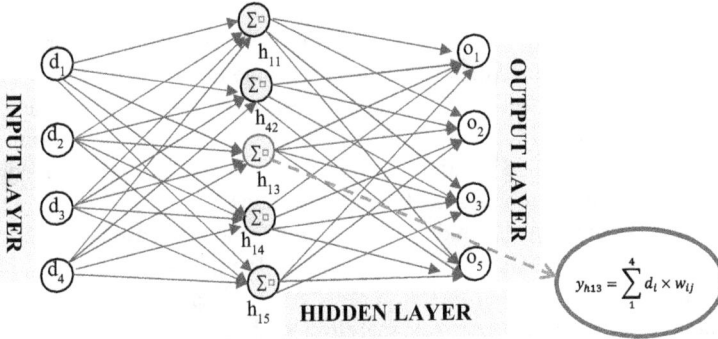

FIGURE 5.5 Typical operational structure of ANN. An ANN structure has an input layer, a hidden layer, and an output layer. The input layer handles the training data, testing data, and the new data. A hidden layer is made up of neurons that contain activation functions and act as a summing junction. Activation functions (weights) are continuously adjusted depending on the input data. Adjustments continue until the desired output is achieved at the output layer. AI models vary in the way that they handle hidden layer operations.

Training the networks iteratively seeks a minimum loss or cost function. Repetition training runs may be necessary to adjust the weights for the expected results. The training process is repeated over numerous sets of training samples until learning is completed by obtaining a minimized loss function. Then, the network is ready for the test data.[E26]

ANN has different methods of learning, some of which are given as follows:

- *Backpropagation neural network* where the errors between input and output are compared for retraining purposes. This is a simple but common ANN technique.
- *CNNs* strengthen the connection between neurons that are "close" to each other by convolving (rolling or entwining) locally.
- *DL network* uses multiple layers having thousands of neurons.
- *FNN* where the flow takes place only in one direction.
- *Fine-tuning network* is a base model in which a few final layers on newly tagged data are custom-trained without modifying the weights of the other layers.
- *Perception network* uses only one layer for training.
- *RNNs* feed the output signal back into the input, which allows short-term memories to be used from the previous input events.

A common ANN is based on the back propagation technique. This model corrects the errors between the input and the expected output. The input data are taken from the labeled training data. The output of each layer is transferred into the next layer as input for the adjustments of the weights of that layer. The process is repeated until the final output layer is reached. The generated output values are compared with the expected outputs. The differences in error between predictions and obtained values are used for updating weights all over again until an error threshold is reached.[E27]

ANNs are suitable for discovering unknown parameters in the data obtained from experiments on a system. Parameter identification is realized by error minimization techniques, in which the distance between parameterized model predictions and observed data is minimized. ANN models can capture complex nonlinearities in data and improve accuracy and stability. ANNs are used in many calibration-related applications, which include multimedia pattern classification, pattern recognition, biometric identification, robotics, and communications.

Calibration methods also include the least-squares methods, pseudo-inverse methods, LM algorithms, and neural network methods. In many cases, an error model is established first, and then the model parameters are identified. Algorithms based on NN do not introduce model errors; therefore, high accuracy can be expected.[E28]

ANNs are also used in preparing training data for sampling-based sensitivity analysis. This method gives valuable information about the importance of observations and significance of each parameter on the behavior of the system. Comparing the datafitting capability of traditional methods, ANN tends to give superior results. Traditional techniques include classical data interpolation, the least squares, polynomial fitting, Monte Carlo methods, and others. Monte Carlo techniques evaluate the uncertainty of measurements and minimize the linearity errors in device readings. The advantages of ANN include simple implementation and high versatility in approximating nonlinear relationships.[E29] ANN models require a reasonable amount of training and testing data.[E30]

5.7   AUTOREGRESSIVE NEURAL NETWORKS

Autoregression is a statistical method used mainly in time series analysis. This method relies on the past values of a time series function to predict the current and future values. It uses mathematical techniques to determine the probabilistic correlation between elements in a sequence to guess the next element. Figure 5.6 illustrates the structure of an autoregressive NN model. The past values of a time series variable are used to predict the future values.

Autoregressive models find applications in generative AI (GAI), image synthesis, time series predictions, and data augmentation. Autoregressive techniques are

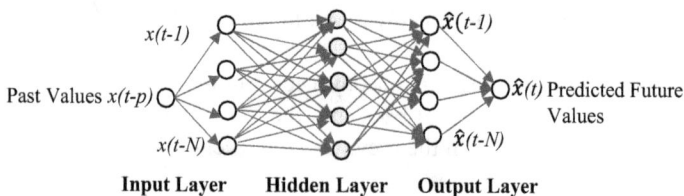

FIGURE 5.6   An autoregressive neural network. An autoregressive neural network predicts the next component in a sequence by taking information from the previous inputs in the sequence. It is used in time sequence analysis by assuming that the current value is a function of the past value. Autoregressive models are used in GAI to create new content learned from the past that can be used for data augmentation, time series predictions, image synthesis, language processing, and so on.

important components of LLMs such as the generative pre-trained transformer (GPT). Autoregressive models use linear regression analysis to predict the next sequence from a known range of variables. In linear regression analysis, statistical models are used to predict the dependent variable from the independent variable.[E31]

Accuracy of calibration relies on the effectiveness of calibration methods. Models such as the autoregressive NN, other DL models, and hybrid models combine the strength of physics-based autoregressive NN models and data-driven models. Autoregressive NNs are generalizable and can be applied to the parameter calibration of any physics-based system, yielding a good degree of accuracy. Physics-based models are constructed using mathematical equations with limited parameter sets.[E32]

5.8 CALIBRATION OF THE AI NETWORKS

AI models have been demonstrated to give accurate and well-calibrated predictions in complex engineering applications such as communication systems. However, the internal operations of models needed to be calibrated for obtaining good regression and inference. When ground-truth and empirical data applications are involved, internally uncalibrated AI models tend to give unreliable results. Therefore, the calibration of the AI model itself bears special importance to generate credible results for the calibration of physical systems and devices. There are many techniques for self-calibration methods unique to each AI model. Only a few typical examples will be provided here.

AI models report a confidence measure associated with each prediction, which reflects the self-evaluation of the model on the accuracy level of a decision. The models implement probabilistic predictors that produce a probability distribution across all possible values of the output variables. However, the self-reported model confidence may not be a reliable measure of the true, unknown accuracy of a prediction. In such cases, the AI model is said to be poorly calibrated.[E33]

Inference of computational model parameters from empirical data is termed *model calibration*. Model calibration aims to obtain model parameters that are theoretically plausible and can generate model predictions for a good fit with observations. In data-driven applications, the inferred model parameters represent physical quantities that are not directly observed or unobservable at all. Some of these unobserved parameters may not be generated by sensors. Therefore, the inference of physics-based model parameters allows the user to understand the underlying reasons for a discrepancy between physics-based model predictions and observations. This discrepancy is called the *reality gap*.[E34] [E35]

Examples of AI self-calibration are given below, but these examples are not exhaustive.

Spiking neuron network (SNN) models are used to incorporate time into their operational procedure. SNN is an ANN that mimics the leverage timing of discrete spikes as the information carrier. In many applications, natural ANNs can be converted to SNNs. During the conversion process, SNN cannot fully mimic the activation outputs of their ANN counterparts within a limited number of timesteps. This can result in significant additional residual information due to residual membrane potential, thus requiring recalibration of the model.[E36] [E37]

Bayesian neural networks (BNNs) are other examples that can pose self-calibration problems. BNNs are known to produce overconfident decisions, especially from limited data. This problem of overconfidence can be solved by concepts of traditional frequentist DNNs. However, in hardware applications of BNNs, the conventional hardware needs of BNN models are resource-intensive, requiring substantial random number generators for synaptic sampling, which can raise network calibration issues.

Exact Bayesian learning offers formal guarantees of calibration only under the assumption that the model is well-specified. This means that the neural network models should have sufficient capacity to represent the ground-truth data generation mechanism. A model trained with Bayesian learning produces probabilistic predictions that are averaged over the trained model parameter distribution. This assembling approach for prediction ensures that disagreements among models that fit the training data equally are accounted for, thus improving the model calibration. Another family of methods aims at enhancing the calibration of probabilistic models implementing validation-based post-processing phases.[E38]

DL models tend to produce either overconfident decisions or calibration levels that rely strongly on the ground-truth and input patterns. However, over-parameterized DNNs are powerful tools for many prediction tasks involving complex input patterns. It is important to capture accurate quantification of prediction uncertainty in many real-world decision-making applications. Techniques such as Bayesian DL can provide a framework for uncertainty estimation of the distributions of the parameters. A reliable predictive model should be accurate when it is confident about predictions and indicate high uncertainty when it is inaccurate. Overconfidence of DNNs can result in unacceptable consequences in safety-critical applications involving ground-truth, such as in critical calibration systems, medical diagnosis, and autonomous driving.

An effective method in self-calibration of the models is the *dropout*. Dropout has been widely used to solve the overfitting problem in uncertainty estimations. Dropout methods include MC dropout, Bernoulli dropout, and Gaussian dropout. However, dropout-based methods require many repeated feed-forward calculations with high computational costs.[E39]

Leveraging implicit or explicit regularization during the training of DNNs can make better calibrated predictions by avoiding overconfidence. One method is the post-hoc calibration, which addresses the miscalibration problem by equipping a given neural network with an additional parameterized calibration component that can be tuned with a validation procedure. There are various techniques to solve over-fitting, such as label smoothing, norm in function space, and focal loss.[E40]

CNNs are typical examples of AI models that are mostly developed with complex architectures to enhance learning capacity with self-calibration features. Self-calibrated convolution is an efficient way to help convolutional networks learn discriminative representations by augmenting the basic convolutional transformation per layer.

CNNs employ multiple filters at each convolution layer to capture a wide array of features from the feature map. Multiplicity and the resultant depth are central to the network's success in processing information in a comprehensive and nuanced form. Each filter detects different features in the input, such as colors, edges, textures, and complex shapes.[E41]

5.9 BPNN AND FNN

FNNs are a foundation concept of neural networks and DL. In an FNN, the information moves in only a forward direction, from input nodes to output nodes through hidden nodes. The activation functions are nonlinear, thus enabling the network to adapt and learn complex data. Common activation functions are sigmoid and Tanh functions, as shown in Equations 5.9.1 and 5.9.2.

FNN does not contain any cycles or loops in the network, unlike RNNs and CNNs. Once a prediction is made, the difference between the predicted output and the actual output is calculated. This error is then propagated back through the network, and the weights are readjusted to minimize the difference between the predicted output and actual output by using optimization algorithms. Popular optimization algorithms are the gradient decent optimization, batch gradient descent, SGD, and mini-batch gradient descent.[E42]

Sigmoid function

$$\sigma(x) = \frac{1}{1 + e^{-x}} \tag{5.9.1}$$

Tanh function

$$\tanh(x) = \frac{e^x - e^{-x}}{e^x + e^{-x}} \tag{5.9.2}$$

Where

e = base of natural logarithms, and

x = numerical value to be transformed.

FFNs find applications in pattern recognition, time series analysis, classifications, regression analysis, and image recognition. A typical example of the use of FNN is compensation of the measurements of environmental factors and decision-making for compensation.[E43]

BPNNs (colloquially termed as vanilla networks) have high learning ability and are used extensively in calibrations. They are versatile and easy to use. A three-layer BPNN can fit any nonlinear continuous function. BPNN is a feedforward neural network that adjusts weights in the hidden layers while training. It uses error back-propagation to improve its performance, as shown in Figure 5.7.

In BPNN, errors between the input and output are compared for retraining purposes. There are numerous effective algorithms for training BPNNs, including gradient descent, Gauss–Newton, momentum, adaptive learning rate, Levenberg–Marquardt algorithm, and so on. For example, the Levenberg–Marquardt back-propagation (LMBP) method determines the optimal connection weights and biases through the backpropagation procedure. LMBP training algorithms can achieve faster training and convergence speeds when compared to stochastic gradient algorithms, particularly for networks with small structures.[E44]

The structures of the BPNN are the main factors affecting their good or poor performances. Too many hidden nodes can cause overfitting, while too few hidden nodes can result in deteriorating prediction ability. As a rule, the number of hidden

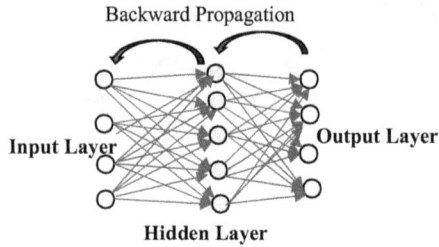

FIGURE 5.7 A BPNN. A BPNN is a feedforward network that adjusts weights through error propagation. It uses backpropagation algorithms to train the networks. The network is trained as a supervised learning method with known inputs and outputs. It aims to minimize the error between the actual output and the predicted output. The error (loss function) between the predictions and desired output propagates backward, and the gradient of the loss function is calculated with respect to the weights. The weights are updated based on the gradient. The BPNN is a popular method for finding applications such as image classification and speech recognition.

nodes of the BP neural network should be as small as possible under the condition of satisfying the requirements of network accuracy. The number of hidden nodes of the BP neural network is usually decided using some empirical formula, so it is difficult to ensure the optimal network structure in each application for efficient operation.[E45]

MLP: DNNs with at least three or more layers are also called MLPs. MLPs are types of FFN in which information flows in one direction from input to output. MLP algorithms use continuous activation functions such as Sigmoid or Tanh. Or they can also use linear, non-linear, and step functions for activations (such as the Heaviside step function).[E46]

5.10　BNNS IN CALIBRATIONS

Bayesian neural networks (BNNs) are an extension of the standard neural network with posterior inference for controlling overfitting. BNNs use statistical methods in determining weights and biases in the model, as opposed to DNNs, in which fixed weights and biases are used in each layer. BNNs treat the weights and biases as random variables, each with its own probability distribution. Initially, the BNN has a prior belief about the weights and biases representing the prior distribution. When data are introduced, prior knowledge is updated, resulting in a posterior distribution, which reflects the knowledge of the model after having observed the data. Bayesian networks perform well in situations in applications with limited data and high uncertainty. They incorporate prior knowledge to learn distribution over possible networks.[E47]

The mathematical explanation is as follows: the likelihood function $p(y|x \cdot \theta)$ representing the probability of observing y output for x input and parameters θ. Posterior distribution representing the probability distribution of parameters, given the observed data, becomes $p(\theta|D)$. The posterior predictive distribution is $p(y^*|x^* \cdot \theta)$ representing the probability distribution of the new output y^*, given the new input x^* and the observed data D.

BNNs are used for parameter description and uncertainty determinations. The uncertainty in observations is used for estimation of the posterior probabilistic description of identified parameters to be compared with the prior expert knowledge. The unknown parameters are modeled as random variables originally supported by prior expert-based probability density functions and updated using the observations to the posterior density functions. BNNs are used in medical applications, autonomous vehicles, active AI learning, and applications containing anomalies and faults.[E48]

5.11   BRAIN-INSPIRED HDC IN CALIBRATIONS

Brain-inspired HDC is an ML application that mimics the high-dimensional data processing capability of the human brain. It uses large vectors to encode information and to perform tasks of classification and recognition. Because of its classification and recognition capabilities, brain-inspired HDC is used in massive sensor networks such as pollution monitoring applications. Brain-inspired HDC is an emerging ML method with great potential compared to traditional ML approaches. Brain-inspired HDC has successfully been applied in voice recognition, gesture identification, image classification, pattern recognition, circuit recognition, large sensor applications, complex hardware design, robotics, Edge AI, and others. Brain-inspired HDC is based on the concept of hypervectors, which are numbers that represent a point in space as vectors having thousands or even millions of dimensions. The hypervectors can be processed with integers, real numbers, or symbols.[E42] [E49]

5.12   CNN IN CALIBRATIONS

CNNs are like deep FNNs (Figure 5.8). They have a weight sharing structure that simplifies the complexity of network models and reduces the number of weights. In traditional algorithms, different methods are used to extract features, but CNN learns and filters by itself. This means the network is self-sufficient to automatically extract features.

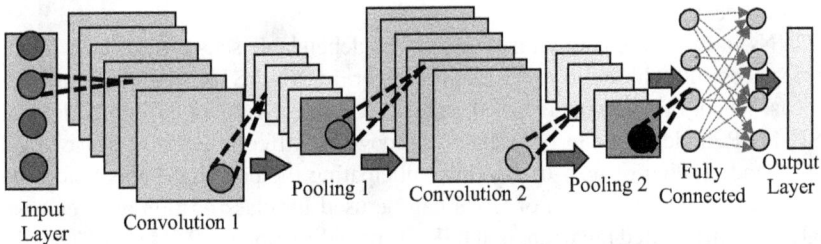

Input Layer   Convolution 1   Pooling 1   Convolution 2   Pooling 2   Fully Connected   Output Layer

FIGURE 5.8   Operation of a CNN. A CNN is a feedforward network that learns features via filter optimization using basic principles from linear algebra. The CNN consists of multiple layers, called a convolution layer, pooling layers, and fully connected layer. It requires large data for training. The CNN is suitable for image recognition, image classification, and natural language processing. (Source: [E50] H. Eren, *Artificial Intelligence in Wireless Sensors and Instruments: Networks and Applications*, CRC Press, 2025).

CNNs employ convolutional layers known as pooling layers (or subsampling layers). Pooling layers serve to reduce the dimensions of the feature maps, lowering the computation needs and minimizing overfitting by retaining the relevant information. The final stage layer is fully connected to execute reasoning based on the learned features, resulting in efficient classifications and predictions. CNNs are used in computer vision, pattern recognition in image processing, speech recognition, and other applications.

A CNN automatically detects important features without any human intervention. The layers apply a convolution operation to the inputs, passing the result to the next layer. The pooling layer reduces the data dimensions by combining the outputs of the neuron clusters. The fully connected layers compute the class scores, resulting in efficient classifications. The four main components of CNN are the convolution layer, rectified linear unit (ReLU) layer, pooling layer, and fully connected layer.

The convolution layer is the core building block of CNN. The convolution layer applies several filters to the input. Each filter activates certain features from the input, such as the edge of an image. The *ReLU layer* is applied to introduce nonlinearity into the model, allowing it to learn more complex patterns. The *pooling layer* reduces the size of the representation, decreasing the number of parameters and computations. This way, it controls and reduces overfitting. The *fully connected layer* is at the end of the network. It maps the learned features to the final output, such as the classing in the classification task.[E51] [E52]

CNNs find extensive applications in calibrations of linear and nonlinear systems, ranging from fault identification to dynamic systems. For example, the CNN is used to identify the parameters of complex nonlinear models with reasonably accurate approximations. In calibration applications, CNN models are capable of learning complex features in measurement hysteresis curves and in other nonlinearities.[E53]

CNN methods are useful in combining statistical optimization and ML methods to obtain accurate and efficient estimates of complex systems. Combined models can recognize complex input features and generate trained, simplified CNN-based surrogate models. The capability of CNN in estimating nonlinear dynamic systems offers significant advantages in terms of calculation accuracy, efficiency, and scalability in measurements.[E54] Some examples of combining CNN with other AI models are explained next.

CNNs in calibrations rely on accurate and dependable sensor data. One method for a successful calibration is the use of a hybrid CNN-LSTM, which is trained to forecast sensor measurements based on historical data. Another hybrid model is the CNN-MLP model that can be trained to recognize different types of sensor faults, biases, and random or poly-drifts, thus anticipating the potential faults before they occur. Various combinations of CNN can be used for classification and regression analyses such as gated recurrent unit (GRU) and bidirectional LSTM (BiLSTM). The CNN-GRU and CNN-BiLSTM have been proven to be highly effective in calibration applications. The performance of these methods is compared using conventional testing methods like RMSE, MSE, and MAE. For the classifications, CNN-MLP has been shown to be very effective in the classification of calibration data.[E55]

In the fault detection of sensors, CNNs integrated with evidence theory can determine fault classifications backed up with uncertainty estimations. Such approaches allow for minimal modifications in the state-of-the-art DNN models by using a risk-calibrated evidential loss function.[E56]

CNNs can also be trained to evaluate the input data in time domains and in time-frequency domains.[E53] [E57] In addition, CNNs are used in dynamic identifications instead of traditional model-based control systems, such as robotic controls. It is worth mentioning that there are other model-free techniques applied in dynamic system identifications, such as multi-layer neural networks (MNN), which utilize architectures based on feed-forward networks (FF-MNN) and cascade-forward networks (CF-MNN).[E58] [E59]

5.13   DL IN CALIBRATIONS

DL is an AI technique that mimics the human brain in understanding and reaching conclusions. DL leads to the development of DNNs, which are advanced forms of ANNs. DNNs can be applied on supervised, semi-supervised, and unsupervised datasets. There is a repertoire of highly successful DL algorithms, such as CNN, RNN, LSTM, and others, as discussed in this book.

Unlike the classical ANN models, DL has many hidden layers (Figure 5.9). Each layer contains multiple neurons acting like summing junctions (adders). Having multiple hidden layers has many advantages in extracting features and learning capacity. The features learned by the DL models can describe the new data with greater accuracy. DL demonstrates the capacity to learn the essential characteristics of data sets from a few samples by applying certain preset rules. It can extract complex features from a large amount of data to discover hidden patterns and trends. The more hidden layers in DNN, the deeper and more intense processing is required for the data to go through.[E60]

It is worth emphasizing that during the training, if the final output differs from the expected output, the weights in the layers are adjusted and the whole process is

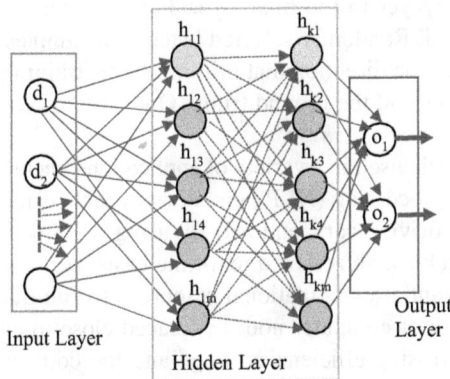

FIGURE 5.9   A deep learning structure. Deep learning models are based on multiple-layer neural networks. The number of layers in the hidden layer can be in the hundreds. A multi-layer structure allows the network to learn complex and abstract representations. Therefore, it finds applications in image recognition, natural language processing, and predictive analytics.

repeated, usually starting from the input layer. In each iteration of training, the amount of deviation between the generated output and the expected output is compared for determining discrepancies. The discrepancies are known as the *loss functions*. The goal is to bring the loss function to zero or close to zero as much as possible. This makes the DL algorithm more accurate in extracting the correct features or patterns from a given dataset. When the loss function is minimized, the network is said to provide the best result it could deliver, and it is ready for a new input.[E51] [E61]

DNN models are extensively used in calibrations when sufficiently large data are available. DNNs can extract complex features from calibration data and discover hidden patterns and trends that are hard to capture by other methods. In many applications, DNNs rely on supervised networks; hence, there is a need for high-quality labeled data for predicting the performance of a system. However, DNNs are also flexible for processing multiple features in unstructured data.[E62] [E63]

Calibration techniques pose significant challenges such as frequency dependence, instrumental effects, environmental influences, drift, aging, interference, and so forth. DL architecture can detect subtle correlations, non-linear dependencies, and higher-order interactions within the data. The capability of DL to extract valuable information from complex datasets makes it a popular choice compared to conventional methods. In many applications, DL-based calibration has demonstrated high accuracy, high R^2, and low RMSE.[E64]

DNN networks in calibrations consist of a feature extraction network, a global regression network, and an output network. A *feature extraction network* extracts sensor modalities since, in many applications, the sensory data are different. In a *global regression network*, the model learns the interrelationship between modalities and output for a concise feature map. *Output layers* give the results of predictions.[E65]

Designing a good DNN model or selecting the right models for a given application may be a challenging task due to the large number of hyperparameters. In ground-truth applications, for instance, multiple passes may be needed over the layers to optimize the parameters of each layer of matrix multiplications one by one, starting from the first layer to the final layer. For this, typically, stochastic gradient algorithms are used. Randomly selected batches of samples can also be used to update the gradients in the direction that minimizes the training loss (the difference between the predictions and the ground truth). One pass through the entire training data set is called a *training epoch*.[E66]

DL is currently widely used in a variety of applications, including computer vision and natural language processing (NLP), end devices such as mobile devices, smartphones, robotics, self-driving cars, and IoT. By using DL models, data generated in these applications can be analyzed on a real-time basis. However, DL inference and training require substantial computational resources to run quickly. Edge computing, where a fine mesh of computer nodes is placed close to end devices, provides some answers in addressing efficiencies. Edge and fog computing are viable ways to meet the high computation and low-latency requirements of DL. Equally, the use of graphic processing units (GPUs) is an important factor in the efficiency of DNN inference and training. Nvidia, for example, provides GPU software libraries to make use of its GPUs, such as CUDA for general GPU processing and cuDNN, which is targeted for DL.[E67] [E68]

A good example of real-time application of DL techniques is on-orbit calibration of microwave and millimeter-wave radiometer spaceborne instruments, including onboard small satellites. The DL calibration model learns the radiometer noise characteristics from radiometer prelaunch measurements during vacuum chamber testing. These pre-learned measurements help to improve the real-time calibration of operating satellites on orbit.[E69]

5.14 DEEP REINFORCEMENT LEARNING (DRL)

RL is an interactive learning paradigm capable of learning so that it can prepare itself for self-improved performance. RL is suitable for supervised and unsupervised learnings by not requiring labeled datasets. RL uses agent theory, and each agent learns and makes suitable decisions to maximize a long-term reward. The desired behavior is rewarded, and the undesired ones are punished (Figure 5.10). Typical RL algorithms include Markov decision process (MDP), Q-learning, policy learning, actor critic (AC), deep RL (DRL), multi-armed bandit, and a few others.[E70]

RL applies biological aspects of learning to solve problems without reliance on the availability of labeled data beforehand. There is no separation between the training and application phases; therefore, the system continuously learns as it predicts. RL trains agents to respond to maximize the values by a trial-and-error method. RL is useful for training computers to play games and training robots to perform tasks. [E71] [E72]

Some of the algorithms employed in RL are given as follows:

- Deterministic policy gradient
- Learning automata
- MDP
- Proximal policy optimization

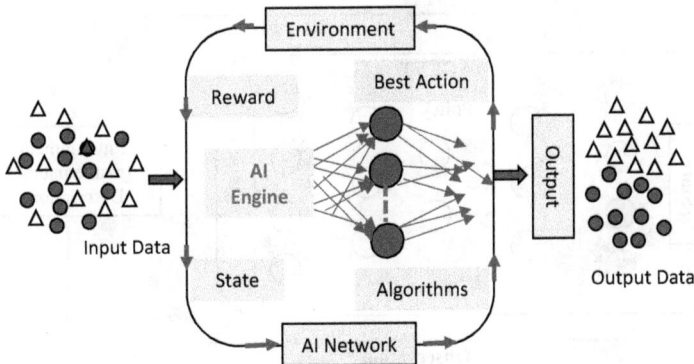

FIGURE 5.10 A typical reinforcement learning method with a feedback loop. RL makes decisions through agents. Agents learn by trial and error to make decisions in a dynamic environment to maximize their rewards. Good actions are rewarded, whereas bad ones are penalized. Actions are a choice that an agent makes in an environment such as video games or driving cars. RL finds applications in robotics, autonomous driving, and game playing.

- Q-learning
- State–action–reward–state–action
- Temporal difference learning
- Trust region policy optimization

DRL combines RL with DL. DRL works well in physics-based models and real-time calibrations. In many applications, DRL learns the calibration strategy directly, guaranteeing robustness and adaptability to changing conditions.[E73]

In many scientific and engineering applications, the dynamic, real-time, and accurate inference of model parameters from empirical data is available. However, accurate inference of these processes with large datasets may not be easily achievable, particularly under noisy conditions. The inference of model parameters with traditional techniques can introduce computational and statistical challenges in determining accuracy. In such situations, DRL comes in handy by forcing the response of the physics-based model to follow the observations. Methods such as Lyapunov-based AC algorithms enable robust and accurate inference of many physics-based model parameters in noisy real-time conditions.[E74]

Lyapunov methods use a scalar Lyapunov function to determine the stability of a dynamic system without the need for system differential equations. In RL, Lyapunov-based AC algorithms contain two main components: (1) an "actor" that determines which action to take in accordance with a policy function and (2) a "critic" that evaluates those actions according to the value function.[E75]

DRL algorithms are primarily based on encouraging positive action through a positive reward and conversely discouraging negative action through a negative reward (Figure 5.11). This allows the agent to adapt to the real-life models by learning quickly. ANN models such as backpropagation enable the DNN-based agent to perform well. An agent with a DNN with learning capacity gets a quick and accurate weight update based on the DRL policy. Hence, DRL has become an effective model

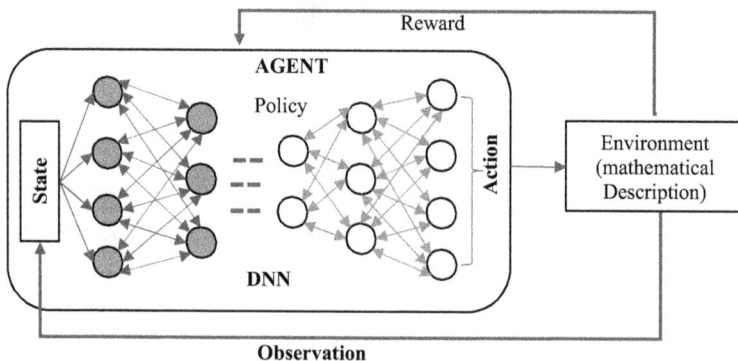

FIGURE 5.11   A deep reinforcement learning. DRL is a combination of deep learning and reinforcement learning. This combination gives more powerful and better decision-making capabilities to the network. Agents digest and interpret complex and high-dimensional data to make decisions autonomously. DRL finds applications in robotics, finance, healthcare, energy management, and other areas.

for learning by mimicking the decision flows of highly complex real-life models. A typical example is in intelligent automation systems in which many sensors are used, requiring timely, efficient, and accurate calibrations. Another example is camera calibration, in which internal and external parameter calibration are divided into groups to adjust the parameters of the value function. The convergence of the value function is calculated to optimize RL.[E76] [E77]

Improved versions of DRL are parallel RL and BiLSTM networks employed with DRL to capture interaction between multiagents. This interaction can be modeled as a discrete discounted MDP, and a learning agent interacts with a stochastic environment in the RL framework. This approach has been shown to work well in cyber-physical systems such as the hybrid tracked vehicles.[E78] The agent interacts with dynamic models to learn tracking and calibration strategies. The calibration can be achieved by an MDP, which does not require any labeled data or ground truth information. MDP is a mathematical modeling of sequential decision-making for optimal policies for maximizing expected rewards over time. MDP is a stochastic control process involving randomness and partial controllability.

To evaluate the performance of the methods, linear regression, the Bland–Altman plot, and the RMSE can be selected for testing experimental data. RNN-based models can replace the entire measurement estimation process, including interpolations and corrections.[E79] [E80]

5.14.1   Q-LEARNING

Q-learning is a model-free, value-based, off-policy algorithm developed from RL. It finds the best series of actions to take, based on the current state. The "Q" stands for quality. Quality represents how valuable an action is to maximize future rewards. Q-learning is basically a method that learns by experience. Q-learning is often combined with other DNNs.[E81] In an application, Q-learning is used with CNN to extract features from video frames for controlling robots.

Bellman's equations are often used in Q-learning. Bellman's formula is a recursive method for optimal decision-making. The formula helps to calculate the value of a given state and to determine its relative position. The state with the highest value is the optimal state.

5.15   FL IN CALIBRATIONS

FL is a distributed but collaborative learning method. The data are trained in a shared and synchronized format by multiple devices and servers. In some cases, the sharing of data is organized by a central coordinator (Figure 5.12), thus avoiding the need to process all the data in a central location. The challenges in FL are the resource allocation, privacy issues, communication costs, and security problems that exist in all distributed systems. FL is used in large projects such as the Deep Space Surveillance Systems.[E82]

FL is widely used in calibration, particularly in distributed measurements and sensing applications. Despite its wide applications, FL has several operational shortfalls, such as the user data privacy and overcoming the limitations due to a single point of

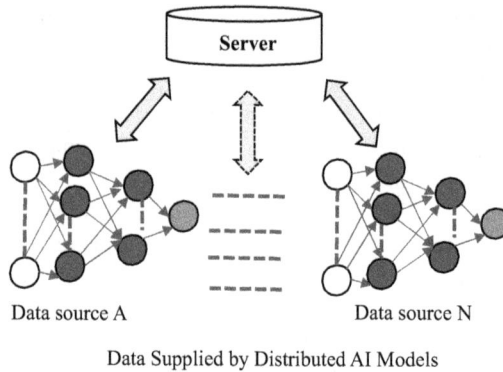

Data Supplied by Distributed AI Models

FIGURE 5.12 A federated learning structure. In FL, multiple devices collaboratively operate to train a model without exchanging raw data. Each device trains a local model and updates the central server. The decentralized operation avoids transmission of local data, thus preventing data losses and breaches in data security. FL is used in mobile devices, IoTs, healthcare, and finance. They are used in the calibration of geographically dispersed devices.

failure. Integration of blockchain technology into FL systems improves some of these shortfalls and partially addresses security, fairness, and scalability.[E83] Methods such as Federated Human Activity Recognition have potential in privacy protection by collaboratively learning global activity recognition through unimodal or multimodal data distribution.[E84] Some methods allow for the use of blockchain-empowered FL framework such that the learning is possible entirely in a decentralized manner.[E85]

Existing FL algorithms can be categorized into horizontal FL, vertical FL, and federated transfer learning algorithms. *Horizontal FL* refers to the cases in which each party has different samples, but the samples share the same feature space. A training step is decomposed to compute optimization updates on each client and then aggregate them on a centralized server without knowing the clients' data. *Vertical FL* refers to the setting in which all parties share the same sample space, but different features use homomorphic encryption for vertical logistic regression-based model learning. The *federated transfer learning* algorithm combines FL and transfer learning. It is an effective collaborative model training without excessive data sharing.[E86] [E87] [E88]

5.16 GAI IN CALIBRATIONS

AI can be considered to have two basic types: (1) predictive AI and (2) GAI. Predictive *AI* aims to generate accurate predictions and forecasts about future events. *GAI* uses most AI techniques, and it focuses on creating new and original contents. This gives AI a degree of intelligence.

GAI finds wide applications in generating images, texts, audio, music, computer codes, videos, and so on. It can produce a variety of novel contents without repeating the trained data. GAI can be constructed by using supervised, unsupervised, and self-supervised ML methods. GAI can be unimodal trained with one type of input, or multimodal that can take multiple forms of data sets such as images, text, and audio simultaneously.[E89]

Types of GAI are given as follows:

- Autoregressive models
- Diffusion models
- Generative adversarial networks
- Generative pretrained transformer
- RNNs
- RL models
- Transformer Base Models
- Variational Autoencoders

Autoregressive models, CNNs, RNNs, and RL models are explained above. GANs, diffusion models, GPTs, VAEs, and others used in calibrations will be explained below.

5.17 GENERATIVE ADVERSARIAL NETWORKS (GANs)

GANs adopt an approach for maximum likelihood estimation and employ two neural networks that compete against each other in a zero-sum game. The GAN architecture (Figure 5.13) utilizes a generator model and a discriminator model. *Generator* models make random noise as input and generate data just like the training data. Generator aims to produce random data indistinguishable from real data by the discriminator. *Discriminators* model takes the real and generated data as input and attempts to distinguish between the two. The discriminator is trained to improve the accuracy of detecting real data vs. generated, while the generator is trained to fool the discriminator. Examples include game development and artificial video generation.[E90]

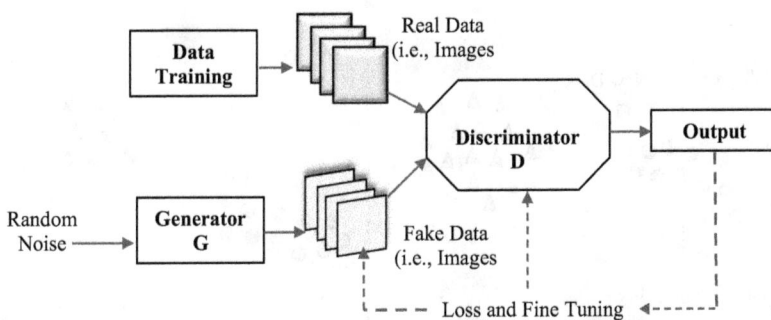

FIGURE 5.13 A GAN structure. GANs train two neural networks to compete against each other. The generator network learns to generate plausible data (fake data), and the discriminator learns to distinguish between the fake data and the real data. The discriminator penalizes the generator for implausible results, forcing the generator to produce undisguised data from the real ones. There are various types of GANs: vanilla GAN, conditional GAN, deep convolutional GAN, and super-resolution GANs are the most used types. GANs are used in image recognition, image processing, and natural language processing.

GANs are used in unsupervised learning as well as supervised learning in a similar manner for both generators and discriminators competing with each other in a zero-sum game. This enables GANs to generate new data with the same statistics as the training set, such as the generation of images that look authentic to human observers.[E91]

GANs are one of the most sophisticated AI models that find applications in many fields, ranging from privacy protection to the calibration of physical systems.[E92] In an application, an EEGANet based on GAN has been produced as a data-driven assistive tool for ocular artifacts removal. EEGANet generates multi-channel EEG signals using the EEG eye artifact dataset that contains a significant degree of data fluctuation. GAN is capable of data-driven artifact removal of multivariate time-series bio-signals.[E93]

5.18 KNN IN CALIBRATIONS

KNN is a nonparametric algorithm utilized for both classification and regression. In this method, the algorithm aims to group the data into different classes. The classes are determined by a popular vote of the nearest K neighbors of an object. The KNN algorithms are suited to multi-modal classes pertaining to many class labels. The error rate of KNN can be higher compared to other methods since the irrelevant features of data during classification can significantly contribute to the outcomes. A typical structure of KNN is shown in Figure 5.14.[E94]

KNN considers the surrounding neighborhood as an input to obtain information about the output (Figure 5.14). This is in contrast with many calibration techniques that use the entire dataset in the prediction process.

In clustering algorithms, KNN uses the Euclidean distance formula to find the similarity between the data points. The distance between the data points x and y can be determined by:

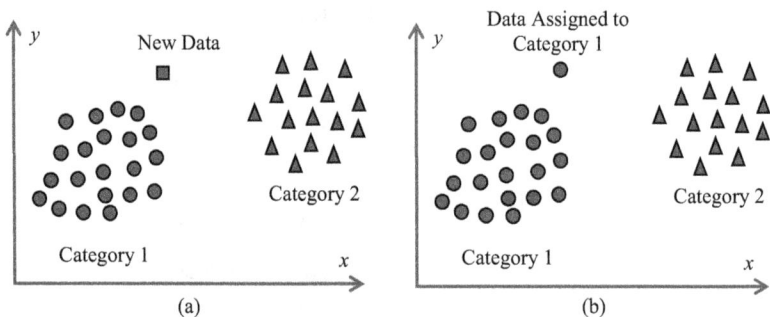

FIGURE 5.14 An operation of a KNN method. KNN is a simple supervised NN used in classification and clustering. In classification, a new data point (a) is placed near the nearest neighbor in the feature space. The algorithm finds the most similar data points (b) that appear frequently among those neighbors from the dataset. It uses the Euclidean distance formula for distance calculations. The number "k" of nearest neighbors needs to be defined in the beginning.

$$d(x,y) = \sqrt{\sum_{i=1}^{N}(x_i - y_i)^2} = \sqrt{\sum_{i=1}^{N}(x_1 - y_1)^2 + (x_2 - y_2)^2 + \cdots + (x_N - y_N)^2}$$

(5.18.1)

Apart from the Euclidean distance formula, there are other methods based on multivariate techniques for distance calculations.

In an application, KNN is used as a two-step hybrid calibration model with SVN. A combination of KNN and SVN is implemented by applying KNN to recalibrate SVN measurement and predicted measurements.[E95]

5.18.1 K-MEANS CLASSIFICATION AND CLUSTERING

The K-means classification and clustering algorithm classifies an object into K-number of objects based on the features and attributes. Grouping is carried out by reducing the number of squares of distances between the data and the corresponding centroid cluster, as shown in Figure 5.15. This technique uses the Euclidean distance metric to divide n observations into k clusters. Its main objective is to minimize the variance (sum of square distances) between the data and the corresponding centroid of the cluster. K-means clustering is useful for feature learning and for providing a starting point for other algorithms.[E22]

K-means clustering assumes spherical clusters that are separable from each other in a way that the mean value converges toward the center of the cluster. The clusters are expected to be of similar size, so that the assignment to the nearest cluster center can be correctly done.[E96]

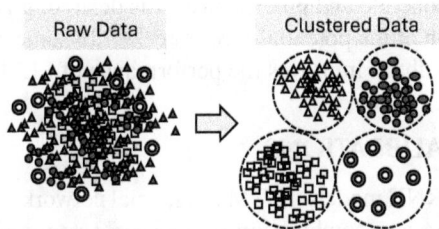

FIGURE 5.15 K-means clustering output. K-means clustering groups data points into clusters based on their similarity. Data are partitioned into k clusters, and each point is assigned to the nearest cluster. Each time a datapoint joins a cluster, the centroid is adjusted until the cluster is stabilized. The number of "k" clusters needs to be defined in the beginning. Initially, clusters are assigned randomly with random centroids. Then, each data point is assigned to the nearest centroid until the cluster assignments do not change. The distance between the data points and centroids is commonly calculated by the Euclidean distance formula. K-means clustering is an unsupervised algorithm; hence, it does not need labeled data training.

5.19   LLP AND LLM IN CALIBRATIONS

NLP enables computers to perform a wide range of natural language-related tasks, ranging from parsing and part-of-speech tagging to machine translations and human dialogues. NLP uses various techniques such as CNNs, RNNs, LSTM, and trans-formers. Based on these models, LLMs can be constructed in multiple natural languages to translate one language to another. LLMs can analyze grammar, summarize text, express sentiments, and generate non-repeat texts. Language models are usually trained on very large language datasets.

ML approaches for language processing have been based on SVM and logis-tic regression for training. Neural networks based on dense vector representations proved to be superior in NLP processing. This trend has sparked applications of DL methods. DL enables multi-level automatic feature representation learning.[E97] [E98]

NLPs and LLMs are difficult to use in direct measuring device calibrations due to their text-to-text nature of operations. However, they are used as assisting tools in physical calibrations. For example, LLMs are applied in cellular networks to har-ness collective intelligence for efficient network management by on-device LLMs in a multi-agent system architecture. On-device deployments suggest personalization of LLMs for better alignment with human intent. Toward personalized generative services, a collaborative cloud-edge methodology becomes feasible, as it facilitates the effective orchestration of distributed communication and computing resources. Integration of communications and computing resources and careful calibration of logical AI workflow leads to intelligent network management.[E99]

Although some LLMs provide internal probability scores for generated tokens, they themselves may be poorly calibrated for indicating the true error rate. Various techniques have been developed for model calibration, including isotonic regression, Bayesian binning, and so on. Some methods rely on annotated calibration data and access to the output probabilities of models. The text-in-text-out nature of the gen-erative language models makes it challenging to quantitatively estimate the error probability. When a concrete and precise answer is desired, it is important to have a quantitative estimation of the potential error rate.[E100] In some cases, the application of RL with human feedback improves the performance of LLMs.[E101]

5.20   LSTM IN CALIBRATIONS

LSTM is a version of RNN models. LSTM is a special network structure (Figure 5.16) with "gates." The gate is a combination of a sigmoid and a bitwise multiplication operation. The sigmoid is an activation function that outputs a value between 0 and 1 describing how much information of the current input can pass through this struc-ture. Although LSTM introduces more complexity in training, it works well in large datasets and in long sequences.[E102] [E103]

LSTM is used to solve the vanishing gradient problems of the conventional RNN models since it is capable of learning long-term dependencies in the data. A typical LSTM cell comprises an input gate, an output gate, and a forget gate. The input gate controls the input flow entering the cell, whereas the output gate controls the output flow of the cell. The forget gate determines the data that should be erased for the

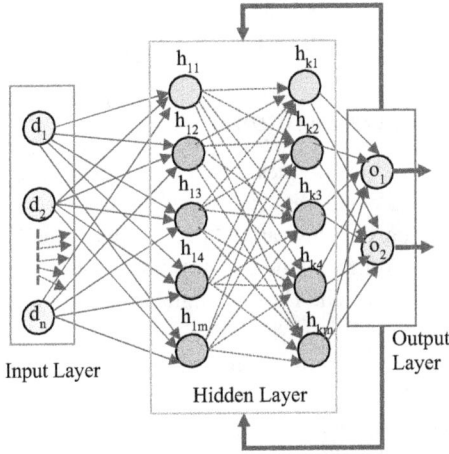

FIGURE 5.16 An LSTM with gates and hidden layers. LSTM is a type of RNN. It can retain long-term dependencies in sequential data. They use memory cells and gates for retaining or discarding the information at hand. The output is fed back to update the hidden layer. The use of previous updates in the forget gate gives LSTM short-term memory capabilities. LSTMs provide solutions to overcome the vanishing gradient problem that exists in RNNs. They find applications in time series analysis, text, and speech applications.

LSTM cell. LSTM has a high computational complexity but can achieve high accuracy, thus easing complexity in large physics-based model calibrations.[E104]

5.21 OPTIMIZATION IN CALIBRATIONS

Optimization techniques are essential in AI training and learning models. Optimization helps minimize the loss functions and improve convergence rates. There are relatively new optimization methods, such as adaptive and metaheuristic methods. As the complexity of AI models increases, effective use of optimization plays a key role for improving models.[E105] [E106] [E107]

Optimization is an established area that has been in use for decades. Therefore, there is a rich repertoire of optimization techniques. Optimization techniques can be grouped as first-order optimization, second-order optimization, and heuristic and metaheuristic optimization.

5.21.1 FIRST-ORDER OPTIMIZATION

First-order optimization techniques rely on gradient information. The key algorithms often used in AI applications are: *Gradient Descent*, which minimizes the loss function by iteratively updating weights of the nodes in the direction of negative gradient. *SGD* updates weights by randomly selecting subsets of data. This method improves efficiency in computing. SGD finds wide applications in DL models having large datasets. *Momentum-based models* incorporate and use past gradient information. This method finds wide applications in DL, as it helps escape the local minima.

Adaptive methods dynamically adjust the learning rate for each parameter, thus improving convergence rate and stability.[E108]

5.21.2 SECOND-ORDER OPTIMIZATION

Second-order optimization techniques use Hessian information to refine gradient updates, leading to more accurate convergence. The Hessian matrix is a square matrix of second-order partial derivatives of a scalar-valued function (a function with multiple inputs returning a single output). It describes the local curvature of a function with many variables. The Hessian matrix is used for determining maxima, minima, and saddle points.[E109]

Examples of second-order optimization techniques are the *Newton method,* which uses second-order derivatives for updates. It needs to compute a full Hessian matrix; hence, it can be computationally intensive. The *quasi-Newton method* simplifies and approximates the Hessian matrix to reduce computation needs and maintain efficiency. *The conjugate gradient method* uses a quadratic function itself without using the Hessian matrix. It is used in large-scale AI applications.

5.21.3 HEURISTIC AND METAHEURISTIC OPTIMIZATION

There are many heuristic and metaheuristic optimization techniques, some of which are: *genetic algorithm (GA),* which is a method inspired by natural selection principles. GAs are good at optimizing hyperparameters and model structures. *Particle swarm optimization (PSO)* is a population-based algorithm mimicking social behavior for optimal solutions. *Simulated annealing (SA)* is a probabilistic method that explores the solution space by gradually reducing a "temperature" parameter. *Bayesian optimization* is a probabilistic approach to optimizing hyperparameters based on prior evaluations. *Ant colony optimization (ACO)* is also a bioinspired method used in combinatorial optimization problems, where agents mimic the foraging behavior of an ant colony. *Differential evaluation* is an algorithm that optimizes real valued functions efficiently in high-dimensional spaces. The *global search algorithm (GSA)* aims to find the best solution within the solution space. It uses multiple starting points and heuristics to avoid getting stuck in local optima. GSA is used to address the inefficient convergences caused by high-dimensional parameters. There are many other methods.[E110]

Challenges in optimization in learning are hyperparameter selection, stability issues, convergence to global optima, and robustness against noisy data.

GAs find wider applications in calibrations. They are also known as the function optimization algorithms, based on Charles Darwin's theory of evolution. They operate on principles like the mechanisms of species evolution through genetic combination and alteration. GA is used for formulating and solving complex problems with multiple variables. The algorithm starts by generating an initial random population of individuals, representing potential solutions to the problem, and then proceeds through several generations. In each iteration of the algorithm, the population generation is updated, and the fitness function (objective function) is evaluated to assess the performance of individuals.[E111]

Shuffled complex evaluation (SCE) and PSO have been popular methods in optimization calibration applications. They are found to exhibit advantages over each other for different reasons. The efficiency of PSO is preferable for relatively deterministic objectives, but SCE may be preferable for relatively flexible objectives.[E112]

An example of calibration using optimization methods is in the energy management strategy. The energy management strategy can be classified as a rule-based strategy, optimization-based strategy, and learning-based strategy. *Rule-based strategies* depend on a set of predefined criteria without knowledge of real-world conditions. When sufficient information is known prior, many methods can be opted for the optimal control strategies, such as dynamic programming (DP), stochastic DP, Pontryagin's minimum principle, model predictive control, equivalent consumption minimization strategy, and others.

5.22 PHYSICS-INFORMED NEURAL NETWORKS (PINNs)

PINNs combine physical models with DL methods, as shown in Figure 5.17. They use differential equations to solve problems, combined with the training of neural networks. It is most effective in minimizing the lost functions in physical applications that have data constraints and severe boundary conditions.[E113]

Applications of PINNs offer solutions for difficult and complex physics-based differential equations to deal with noisy and uncertain observation datasets. They

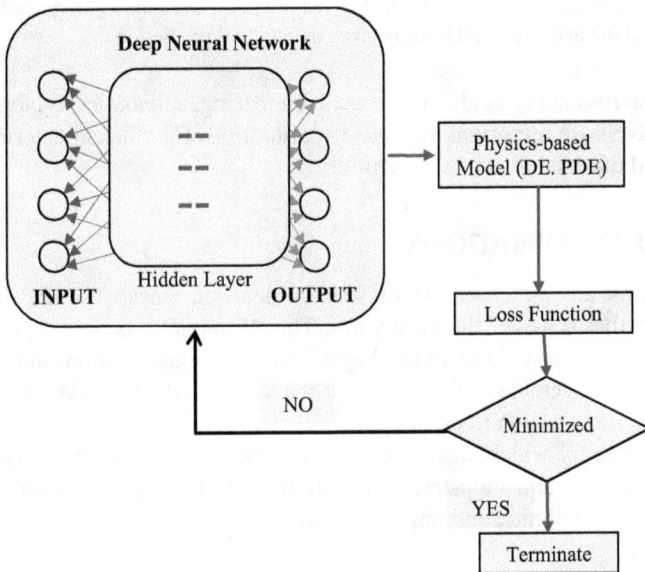

FIGURE 5.17 A physics-informed neural network. The PINN incorporates physical laws and deep learning processes. High-order differential and partial differential equations are typical examples representing the physics laws. The AI model predictions are compared with the predictions of physics solutions to determine the loss function. If the loss function is high, neural network parameters are readjusted until the loss function is minimized.

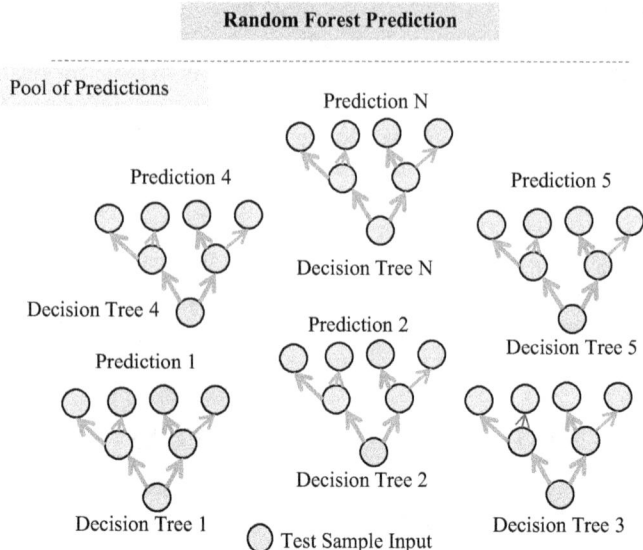

FIGURE 5.18 A random forest structure. RF uses an ensemble of decision trees, each tree making predictions autonomously. It is suitable for classification and regression tasks in supervised learning. RF methods improve accuracy and reduce overfit problems that exist in many AI models. RF trees randomly select a subset of a dataset for training of each tree by using the bootstrap sampling method. The results of all trees are aggregated for the final prediction, and the average of all trees is taken as the final prediction.

are applied across many fields in science, engineering, economics, biology, and life sciences. PINNs are important methods in calibrations for efficient determination of accuracy and uncertainties in a system.[E114]

5.23 RF IN CALIBRATIONS

RF algorithms are an ensemble of sets of decision trees. The RF selects the classification that achieves the most votes. The RF model produces an *ensemble* of randomized decision-making trees (Figure 5.18) for classification and regression. The aggregated ensemble either combines the votes modally or averages the probabilities from the decision trees.[E115]

RF is suitable for widely distributed measurement systems such as IoTs, environmental monitoring, and industrial applications. RF models have been extensively used to monitor, calibrate, and improve air quality.[E116]

5.24 SEMI-SUPERVISED, UNSUPERVISED, AND ACTIVE LEARNING

Training DL models requires a large amount of labeled data. However, in some applications, labeled data are either scarce or totally unavailable. An example is medical image analysis, which is hard to obtain, expensive, and time-consuming. Many applications require the development of label-efficient DL methods, but some other methods are semi-supervised learning (SSL), unsupervised learning, and active learning.

5.24.1 Semi-Supervised Learning

SSL involves human-in-the-loop for reviewing questionable predictions. It applies to supervised learning models from small and tagged datasets to large and untagged datasets. In some applications, the combination of labeled and unlabeled data improves the learning performance. The purpose of SSL is to separate the combination of labeled data and unlabeled data.[E117]

There are several methods in SSL, which are listed as follows:

- Active learning
- Few shots learning
- Co-training
- Generative learning
- Graph-based learning
- Low-density separation learning
- Transduction

SSL addresses the ubiquitous issue of label scarcity. The state-of-the-art SSL methods utilize consistency regularization to learn unlabeled predictions that are invariant to input-level perturbations.[E118] [E119]

5.24.2 Unsupervised Learning

Unsupervised learning models rely on unlabeled data for inferring the underlying information structure without being dependent on external resources or human supervision. The advantage of unsupervised learning is that no prior knowledge is required; however, this comes at the cost of potential reduction in accuracy. Another disadvantage is that the automatically discovered data may not always be representative of real-world situations. Given its unique features, unsupervised learning is suitable for solving the problems of user groups for hybrid multiple access, attack detection of malicious users, and so on.[E120] [E121]

Unsupervised learning analyzes a stream of data and finds patterns and features from unlabeled data. It makes predictions without additional guidance. Unsupervised learning can also be effective in dimension reduction, clustering, exploring the data patterns and structures, and finding groups of similar objects. It can detect outliers and noise in the data.[E122]

Unsupervised learning uses previously learned functionality when new data are introduced for the explanation of the data classes. There are numerous methods in unsupervised learning, which are given as follows:

- *Association rule learning:* Apriori, Eclat, and FP-growth algorithms
- *Auto-encoders*

- *Cluster analysis*: conceptual clustering, expectation maximization, fuzzy clustering, hierarchical clustering, K-means clustering, K-medians, mean shift, and single-linkage clustering
- *Dimensionality reduction*: canonical correlation analysis, dynamic mode decomposition, factor analysis, feature extraction, feature selection, independent component analysis, linear discriminant, multidimensional scaling, non-negative matrix factorization, partial least squares regression, PCA, principal component regression, projection pursuit, Sammon mapping, t-distributed stochastic neighbor embedding
- *Expectation-maximization algorithms*
- *Information bottleneck methods*
- *Vector quantization*

5.24.3 ACTIVE LEARNING

Active learning uses less labeled data for model training. Active learning intends to find representative samples (e.g., a small portion of a large pool of datasets) to be annotated by an oracle (e.g., human annotator), such that a well-supervised learning model can be trained intelligently on the selected annotated samples. Generally, active learning techniques can be divided into three groups: membership query synthesis, stream-based selective sampling, and pool-based sampling.[E123] [E124]

1. *Membership query* synthesis method: Learners can select any unlabeled samples from instance space or generated samples by the learners for annotation.
2. *Stream-based sampling* is selective sampling, sequentially sending data samples to an active learner. An active learner decides whether to annotate or to discard the sample.
3. *Pool-based sampling* method aims to select samples for annotation from a pool of unlabeled samples. It can select more than one sample at a time. Pool-based sampling is the most widely used technique in active learning.

5.25 STATISTICAL AI CALIBRATIONS

Many calibration models suffer from low calibration accuracy caused by unaccounted factors and unobservable data errors. To address these problems, AI-based models combined with statistical methods are found to be effective, for example, in dynamic calibrations. Methods such as extended Kalman filter-incorporated with residual neural network-based calibration have been shown to work well.[E125]

When data from several types of heterogeneous sensors are fused together, many factors influence the statistics of the estimation error, and their combined impact tends to be difficult to predict. Once the system is assembled, experimental validation can be complex and difficult, demanding sufficiently accurate ground truth information.

Deterministic errors, such as scale factors and other parameters, can be pre-calibrated and removed from the measurements directly. However, when additive statistical errors are involved, such as white noise, statistical models become useful. Such processes are referred to as "statistical calibration." Accurate statistical calibration allows for maximizing AI estimation accuracy by pre-removal of the maximum portion of errors from measurements. This also helps to identify misbehaving sensors or measurements via fault detection and exclusion/isolation mechanisms.

Two competing statistical modeling techniques are Allan variance linear regression and the emerging generalized method of wavelet moments.[E126] Allan variance, also known as two-sample variance, is a statistical method used in time series analysis, clocks where accurate timing is critical, oscillators, and frequency stability analysis.

5.26 SVM IN CALIBRATIONS

SVM learning is a method that focuses on maximizing the separation of the margin between classes (vectors). SVM is a well-used supervised ML algorithm for classification, which is used on many calibration applications.[E127]

An SVM finds the straight line that best separates the two groups of points on a plane. It is used as the basis of building optimal classification and regression models. The training process is complex, but once tuned, SVM models provide high accuracy, leading to generalization. Types of SVM is illustrated in Figure 5.19.

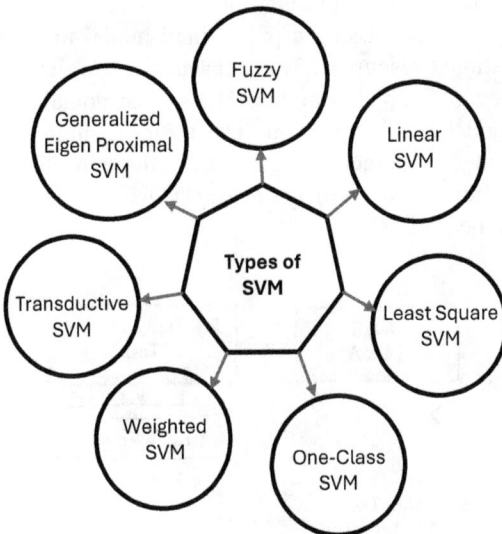

FIGURE 5.19 Types of SVM. SVMs have two main types: linear and nonlinear. If data are linearly separable, they can be separated into two classes by a straight line (called a hyperplane). If data are not separable linearly, kernel functions are used to map the data into higher dimensions where a linear separation is possible. Sigmoid and polynomial functions are used to capture nonlinear relationships. There are numerous SVM techniques to handle nonlinear data.

SVM is a sophisticated supervised ML technique. By using training data, the algorithm enhances an optimal hyper-plane, and this describes the decision made and classifies new samples. SVMs support both linear and nonlinear classifications. In complicated cases, the points can be projected into a higher-dimensional space to find the plane or hyperplane that best separates the classes. The projection is called a *kernel*, and the process is called the *kernel* trick.[E128]

There are numerous options for selecting suitable kernels, some of which are:

Polynomial (homogeneous) kernel $k(x_i - x_j) = (x_i \cdot x_j)^d$ where $k > 0$ and d is the degree of polynomial and $(x_1 \cdot x_j)$ is the dot product of vectors x_i and x_j

Polynomial (inhomogeneous) kernel $k(x_i - x_j) = (x_i \cdot x_j + r)^d$ where $(x_1 \cdot x_j)$ is dot product.

Hyperbolic tangent kernel $k(x_i - x_j) = \tanh(\kappa x_i \cdot x_j + c)$ where $k > 0$ and $c > 0$

Other kernel functions are Gaussian radial basis function, Laplace radial function, Bessel function, ANOVA radial basis function, and linear spline kernel in one dimension.[E129]

5.27   TRANSFER LEARNING

The transfer learning model (teacher–student concept) is trained on one task and reused as a starting point for a model on a related but different task. The knowledge gained in solving a problem on a subject can be applied in related new subjects (Figure 5.20). Using a pre-trained model to train others reduces time and computational resources. Transfer learning minimizes the differences between the source and target domains. The source domain is the dataset that the pre-trained model is derived from. The target domain is the specific task where the knowledge is gained from a source domain. There are two major approaches to minimize domain differences: (1) Distance-based methods and (2) adversarial-based methods.[E130]

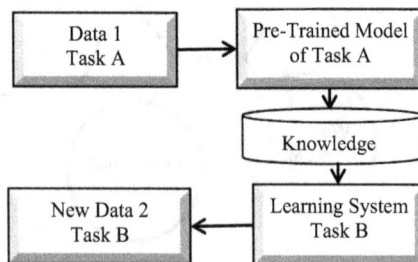

FIGURE 5.20   A transfer learning process. Transfer learning is used often in distributed calibration systems. In transfer learning, a model is pretrained on one task to enhance performance on a related task. Transfer learning retains the fundamental knowledge, thus saving computing time and requiring fewer resources. Types of transfer learning are transductive transfer learning, inductive transfer learning, and unsupervised transfer learning.

Distance-based methods measure and minimize the domain differences such that the gap between the source and target domains is minimal. The source classifier/ regressor is trained using the labeled source domain data to be used for the target domain classification/ regression tasks.

Adversarial-based methods are inspired by GANs.

Transfer learning assumes that the availability of a labeled source domain is related to the target. For example, when performing an image classification task with no available labeled data, an effective way is to use transfer learning from publicly available datasets.

A form of transfer learning is the *fine-tuning* that assumes the availability of a few labeled data in the target domain. Fine-tuning aims to train a few layers of the learned network from the source domain to train by using a few labeled data of the target domain. The source domain can be shared community AI models.[E123]

5.28   AI COMPUTING

AI-related computing platforms can be listed as cloud computing, edge computing, fog computing, and quantum computing.

5.28.1   CLOUD COMPUTING

Cloud computing is a platform that connects many end devices distributed all around the world. The connected devices can be used by consumers, research activities, industry, and business organizations. These interconnected devices generate large amounts of data that require transmission, analysis, and storage. To ensure integration between distributed devices and cloud computing, different interconnection structures and architectures are used.[E131]

The use of AI in cloud computing is widely established at various stages of cloud structure and various applications. For example, the ability of AI-assisted cloud computing to share or offload computationally intensive tasks on remote facilities is an efficient and effective alternative in many applications.[E132] [E133]

End device architectures in a cloud environment have the following layers: (1) connected devices layer, (2) communication layer between connected devices (e.g., Bluetooth and ZigBee), (3) access network layer, (4) cloud layer, and (5) application layer.

5.28.2   EDGE COMPUTING

Edge computing brings information storage and computing capabilities closer to the device producing that information. There are many architectures proposed or adopted in practice to reduce latency and to have fast response times. Some examples are Fog Computing, Mobile Edge Computing, Cloud-Edge-Beneath, Cloudlet, FermtoCloud, Nebula, Mobile-Edge Offloading, and Foraging.[E134] [E135]

AI models are applied in serverless edge computing in runtime mechanisms for scaling, placement, and routing. AI-driven platforms are based on various methods, such as RL for continuous learning and optimizing the functions and deployments.

5.28.3 Fog Computing

Fog computing advocates a user-centric architecture to utilize the context and resources locally. The user-driven shift in computation and storage enables many applications requiring less interaction with remote services (or data centers) for efficient performances.[E136]

5.28.4 Quantum Computing

Quantum computers make use of the quantum mechanical phenomenon of atomic and subatomic particles. They find applications in solving complex optimization problems much faster than classical computers.[E137] [E138] [E139]

5.29 COMPUTING EFFICIENCY IN AI APPLICATIONS

Computing efficiency is a requirement for all types of computing activities. Computer efficiency is necessary in calibrations where a large number of devices are involved, such as the IoT systems, cyber-physical systems, and healthcare applications. One way of increasing computing efficiency is by sharing trained AI models. Sharing can significantly reduce the need for repeating model training efforts. Compressed models, for instance, can be shared on resource-limited devices (e.g., embedded systems), thus requiring less memory and low computational power.[E140]

In computation- and data-efficient learning, a technical challenge is to reduce the sizes of AI models and the amount of training data. For instance, Compressing AI models into smaller sizes without significantly compromising their performance can save time and can reduce the need for resources during the deployment stage. For data-efficient learning, the main objective is to reduce/alleviate the use of large, labeled datasets for model training. The collection of large and labeled datasets requires a significant number of resources.

Computing efficiency using software for AI is achieved by (1) compressed AI modules, (2) pruning, (3) quantization, (4) knowledge distillation, (5) matrix decomposition, and (6) early exit of inference.

5.29.1 Compressed Modules

Compressing modules is an effective way to acquire computational efficiency. Compressed AI models are particularly useful in applications where models are run for routine and long-term tasks, as in the case of wireless communication systems. Compressions can be lossless compressions or lossy compressions. Generally, *lossless compression* reduces the data size for all the data, but it cannot achieve high compression. The *lossy compressions* take place at the expense of losing some data. The original data cannot be reconstructed fully, but they are still applied for efficiency at the expense of some losses.

The compression ratio (CR) provides a direct measure of the compressor's performance, which can be expressed as:

$$CR = \frac{\text{input raw data size}}{\text{Compressed data size}} \qquad (5.29.1)$$

The reconstruction quality in terms of sample-to-sample error can be determined by root-mean-square difference (PRD) and percentage root-mean-square difference normalized (PRDN) as:

$$PRD = 100 \times \sqrt{\frac{\sum \left(x[k] - x_r[k]\right)^2}{\sum \left(x^2[k]\right)}} \qquad (5.29.2)$$

and

$$PRDN = 100 \times \sqrt{\frac{\sum \left(x[k] - x_r[k]\right)^2}{\sum \left(x[k] - \ddot{x}\right)^2}} \qquad (5.29.3)$$

Where
  $x[k]$ is the raw data, and
  $x_r[k]$ is the compressed data.
  The error estimates are expressed in terms of MSE, normalized mean square error (NMSE), RMSE, and normalized RMSE. As an example, NMSE can be determined from:

$$\frac{\sqrt{\sum \left(x[n] - x_r[n]\right)^2}}{\sqrt{\sum \left(x[n]\right)^2}} \qquad (5.29.4)$$

Where for a length of data set N, $x[n]$ = raw sample and $x_r[n]$ = reconstructed sample.[E141]

5.29.2   PRUNING

Pruning is used when AI models are over-parameterized. The network pruning serves to remove redundant weights or neurons that have the least impact on accuracy, resulting in a significant reduction in memory storage and computational cost. Pruning is one of the most effective methods to compress AI models. Figure 5.21 shows the process of pruning for a typical neural network-based AI model.

One of the simplest methods in pruning is to estimate importance in terms of the absolute value of a weight, *called magnitude-based pruning*. In this method, zero-valued weights or weights within a certain threshold are removed. *Regularization* (preventing overfitting by assigning penalties) is a method used often during training to increase weight sparsity. However, magnitude-based pruning methods may result in unstructured network organizations, which may be difficult to compress or accelerate without specialized hardware support.[E142] [E143]

5.29.3   QUANTIZATION

Quantization is another approach for computationally efficient AI. In quantization, floating-point parameters are quantized into lower numerical precision. For example,

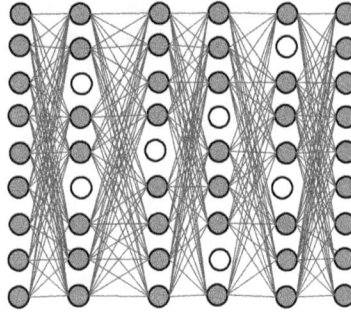

FIGURE 5.21 Pruning eliminates ineffective neurons that have zero (or near-zero) values. Pruning reduces the model size and makes the network faster and efficient. It speeds up the inference process and improves memory usage. There are various types of pruning, including weight pruning, filter pruning, and neuron pruning.

32-bit is reduced to 16-bit, 8-bit, or even 1-bit. Typically, a 32-bit single-precision floating-point format is quantized to 16-bit half-precision float or 8-bit mini float.

Although quantization is generally applied to weights and activations, it can also be extended to optimization and gradient computations. Instead of quantizing a single weight, vector quantization approaches focus on factorizing the weight matrix by employing weight clustering and weight sharing.

A combination of pruning or quantization for network compression leads to higher compression rates. Both pruning and quantization are effective ways to reduce model size for efficient learning.[E144] [E145]

5.29.4 KNOWLEDGE DISTILLATION

Knowledge distillation uses the information from large models (called teachers) to train small models (called students). The student model mimics the teacher model to obtain competitive or even superior performances. Knowledge distillation can be categorized as: logits-based, feature-based, and relation-based.[E146] [E147]

5.29.5 MATRIX DECOMPOSITION

Matrix decomposition relies on modeling parameters of AI models as matrices, for example, describing the weights connecting two layers of a feed forward neural network. The parameter matrices form lower ranks can be decomposed into smaller matrices with an overall reduction in the number of parameters.[E148] [E149]

5.29.6 EARLY EXIT OF INFERENCE

Early exit of inference (EEoI) introduces outputs at different points across a neural network. During the inference process, when an early inference can be performed with high confidence, EEoI allows obtaining a reliable result without reaching the last layers. This method typically incorporates additional trainable layers or a supplementary network, indicating the layers that may be skipped during inference.[E150] [E151]

5.30   CYBERSECURITY

Calibration of devices purely relied on physical interactions and manual adjustments for decades. Today, they are equipped with advanced features such as intelligent systems, wireless network connectivity, data loggers, and remote access facilities. Such advancements provide efficiency and convenience, but they are vulnerable to interference, incomplete or missing data, and cybersecurity risks.[E152]

Wireless communication networks allow for the connection of millions of devices to local and global networks for transmitting and processing data. Wireless systems use wide area networks and local area networks, such as Bluetooth or Wi-Fi. The network connectivity feature enhances the calibration workflow, but it creates vulnerability to potential entry points for cyberattacks. Cyber-attacks can take place during calibrations or, at other times, be activated outside calibrations, which can be serious in sensitive cases such as medical, health, and military applications.[E153]

The aim of network intrusion detection is to discover any unauthorized attempts to a network by analyzing incoming and outgoing network traffic to find out the existence of malicious activities. Traditional approaches, such as supervised network intrusion detection, show effective performance for detecting malicious payloads included in network traffic-based datasets labeled with ground truth.[E154]

Cyber-attacks can cause the following:

- *Manipulation of calibration data* that can compromise the integrity of measurement, leading to false readings.
- *Disruption of calibration processes* by disabling equipment or disrupting calibration workflows. This can cause delays and disruptions in calibration-sensitive operations and industrial production lines.
- *Access to sensitive information* that can interfere with intellectual property issues and the confidentiality of information.

Cybersecurity is an ongoing process. By being informed and implementing the best practices, they can mitigate risks and ensure the safety of calibrations and related devices. Calibration organizations and laboratories protect their equipment and valuable data by prioritizing cybersecurity and implementing strong security measures.[E155] [E156]

AI models are increasingly employed for securing large cloud and edge-based systems. These models require the collection and management of very large amounts of security-related data. The data needs to be efficient and adaptable to different security contexts and deployments. The prerequisite for enabling AI models is the development of scalable infrastructures for collecting and processing security-related datasets from individual connected devices, edge nodes, various application platforms, and from entire cloud.[E157]

AI methods employing GANs with Resnet and Glowworm optimization can detect and prevent both internal and external threats.[E158] In general, standard GANs are found to be effective in cybersecurity issues. A version of GAN is a bidirectional GAN (Bi-GAN) model equipped primarily for network intrusion detection with reduced overheads in training. Other common methods employed in cybersecurity are ANN, RNN, and CNN.

Autoencoders and GANs can generate realistic synthetic data sets to improve detection accuracy based on anomaly detection techniques. An autoencoder is composed of an encoder and a decoder. The encoder can compress high-dimensional input data into low-dimensional latent space. The decoder generates the output that resembles the input by reassembling (reducing) the data representation from the latent space. Typically, a reconstruction loss is computed between the output and the input. The reconstruction loss is then used as a mechanism to identify anomalies like the ones encountered in calibrations.

5.31 AI AND HARDWARE IN CALIBRATIONS

AI-based systems are fueled by increasing computing power provided by edge computing and cloud, as well as the processing of massive volumes of data collected and provided by devices. The combination of big data and AI is used to solve many automation and optimization problems. Integration of AI into the hardware structures comes with many benefits, including efficiency in computing and reliable operations of devices themselves.[E159]

Technology downscaling allows for the integration of more and more systems into a single chip, with the subsequent benefits in terms of increased performance of information storage and management, together with communication and computation facilities, located in end-devices such as mobile phones, distributed sensors, and IoT nodes. Embedding AI modules in systems-on-chip (SoC) is becoming commonplace in mobile terminals where AI engines are used for specific tasks, such as face and fingerprint recognition. AI-based algorithms are also applied to improve the performance metrics of analog circuits by means of linearization or calibration.[E160]

AI-empowered systems benefit from digital signal processing in terms of programmability, scalability, and robustness. The need for earlier digitization makes the analog-to-digital converter (ADC) one of the key components in many SoCs. AI algorithms are applied to improve the performance metrics of ADCs.[E161]

As an example, ANNs have been applied as part of the design methodology for the sizing of analog and mixed-signal circuits, so that they can conveniently be used for the optimization of $\Sigma\Delta$Ms (sigma delta modulators). Optimization-based synthesis methodology is widely used for the automatic sizing of circuits and systems, to find the best set of design parameters that meet the required specifications of a given circuit. AI can be trained to replace simulators.[E162]

AI is implemented in many ways, regardless of whether it is used as an optimization algorithm to design $\Sigma\Delta$Ms or as part of the circuit itself. In the latter case, it is either synthesized in hardware in field-programmable gate array (FPGA) or integrated as an on-chip dedicated module, or software embedded in a digital signal processor.[E163]

Another example is the reconfigurability of $\Sigma\Delta$Ms, which is enhanced by AI algorithms to adapt the specifications of ADCs to diverse input signal requirements, environmental interferences, and noise levels. A high degree of programmability is required, mostly digital/analog circuit techniques as well as suitable topologies of ANNs to implement the AI engine. Moreover, the practical implementation of

AI-assisted $\Sigma\Delta$Ms requires adoption of diverse design strategies—from the $\Sigma\Delta$M architecture itself to AI modules and circuit building blocks.[E164] [E165]

CNNs require a high computational load, indicating parallelization possibility and higher efficiency. GPUs, FPGAs, and application-specific integrated circuits can leverage a higher level of parallelism than conventional processors.

Analog in-memory computing (AIMC) has been utilized in edge inference engines to solve the memory bottleneck problem and increase efficiency. AIMC output quantization is adjusted by controlling analog gain and entangling it with analog parameters and nonlinear functions. AIMC dynamic output quantization is used.[E166]

5.32 EXPLAINABLE AI TRUSTWORTHINESS AND RELIANCE

Important factors in AI are explainability, trustworthiness, reliance, and robustness, which are directly related to applications in the calibrations they are involved in. Other factors discussed in various places in this book are the human-centric features, performance levels, privacy, and security by design.[E167]

5.32.1 EXPLAINABLE AI

The Defense Advanced Research Projects Agency (DARPA) launched a program for eXplainable AI (XAI) in 2017. It aims to shape new learning processes for producing better explainable models, designing effective explanation interfaces, and understanding the psychological requirements for effective explanations.[E168]

Despite the proliferation of ethical frameworks of AI from different organizations such as government agencies, large corporations, and academic institutions, it is still a challenge to implement and operationalize ethical and legal frameworks for AI due to its complexities. The implementation and operationalization involve different aspects in theoretical and practical research on designing, developing, presenting, testing, and evaluating approaches, which are supported by advanced AI techniques. Recently significance progress is made on AI ethics implementation, trust and trustworthiness in AI.[E169]

The theoretical issues in the important aspects of AI trustworthiness, including robustness, generalization, explainability, transparency, reproducibility, fairness, privacy preservation, and accountability, is provided to unify the current but fragmented approaches toward trustworthy AI, ranging from data acquisition and model development to system development, deployment, continuous monitoring, and governance.[E170]

5.32.2 TRUSTWORTHINESS AND RELIANCE

Issues of information trustworthiness are important in AI applications. Reliance is a concept relevant to trust. Trust is attitudinal and a psychological construct, while reliance focuses on the behaviors of humans, which are directly observable, objective measures. Trust in AI involves not only reliance on the system itself but also trust in the developers of the AI system. The literature review on trust and trustworthiness in AI demonstrates the need to consider the wider context of AI system development and use to better understand AI.[E171] [E172]

An example is the verbal Predictive Reliance Calibrator (vPred-RC), an algorithm for selectively providing Reliance Calibration Cues (RCCs) that communicate an AI's capability to users, resulting in optimizing their reliance on it. Understanding what an AI system can or cannot do is necessary for the end-users to be able to use AI properly without being over- or under-reliant on it. RCC is a cognitive model that predicts whether a human will assign a task to an AI agent.[E173] [E174]

Research on AI trustworthiness focuses on the algorithmic properties of models. The success of AI technology has benefited from the accuracy-based performance measurements. By assessing task performance based on quantitative accuracy or loss, training AI models becomes tractable in the sense of optimization. Meanwhile, predictive accuracy is widely adopted to indicate the superiority of an AI product over others. From an industrial perspective, the lifecycle of an AI product consists of multiple stages, including data preparation, algorithmic design, development, and deployment, as well as operations, monitoring, and governance.[E175]

5.32.3 ROBUSTNESS

Robustness refers to the ability of an algorithm or system to deal with execution errors, erroneous inputs, or unseen data. Robustness of AI is essential for at least two reasons: distributional shift and illegal inputs. *Distributional shifts:* The environment where an AI model is deployed becomes more complicated and diverse. If an AI model is trained without considering the diverse distribution possibilities of data generated by different scenarios, its performance might be significantly affected negatively. *Illegal inputs:* Besides vulnerability, the system-level robustness against illegal inputs should also be carefully considered in realistic AI products and diverse applications.[E176] [E177] [E178]

5.33 CONCLUSIONS

The use of AI is making headway in instrumentation and measurements and calibrations. The AI techniques tried and proven to work are explained in this section. DL methods require large data, and the sources necessary for successful calibrations are explained in detail. In addition to traditional calibrations, and due to the availability of cost-effective digital systems and computers, many novel methods are applied in calibrations. Special attention is paid to computing concepts to illustrate how they play important roles in modern measurements and measurement systems.

REFERENCES

E1. S. Martínez-Fernández *et al*, "Software Engineering for AI-Based Systems: A Survey," *ACM Transactions Software Engineering Methodology*, vol. 31, no. 2, Article 37e, p. 1–59, March 2022.

E2. F. Kamalov, S. Moussa and J. A. Reyes, "Data Transformation in Machine Learning: Empirical Analysis," *2023 International Conference on Innovation and Intelligence for Informatics, Computing, and Technologies (3ICT),* Sakheer, Bahrain, pp. 115–120, 2023.

E3. V. Trehan, "AI-Powered Archives: Revolutionizing Information Access for the Future," *2023 IEEE International Conference on Big Data (BigData)*, Sorrento, Italy, pp. 6298–6300, 2023.

E4. L. Huang, J. Zhao, B. Zhu, H. Chen and S. V. Broucke, "An Experimental Investigation of Calibration Techniques for Imbalanced Data," *IEEE Access*, vol. 8, pp. 127343–127352, 2020.

E5. E. S. Maddy, S. A. Boukabara and F. Iturbide-Sanchez, "Assessing the Feasibility of an NWP Satellite Data Assimilation System Entirely Based on AI Techniques," *IEEE Journal of Selected Topics in Applied Earth Observations and Remote Sensing*, vol. 17, pp. 9828–9845, 2024.

E6. Y. Sun, M. Kountouris and J. Zhang, "How to Collaborate: Towards Maximizing the Generalization Performance in Cross-Silo Federated Learning," *IEEE Transactions on Mobile Computing*, vol. 24, no. 4, pp. 3211–3222, April 2025.

E7. National Instruments, *Collaboration Executive*. Calibration Executive Download - NI https://www.ni.com/en/support/downloads/software-products/download.calibration-executive.html#565559 (Accessed on 20 April 2025).

E8. R. Li, X. Wang and S. He, "Two-Stage Generative Color Calibration for Drone Photography with Cloud-Edge Collaboration," *IEEE Internet of Things Journal*, vol. 12, no. 5, pp. 6054–6057, March 2025.

E9. Z. Alshara, A. Shatnawi, H. Eyal-Salman, A. -D. Seriai and M. Shatnawi, "PI-Link: A Ground-Truth Dataset of Links between Pull-Requests and Issues in GitHub," *IEEE Access*, vol. 11, pp. 697–710, 2023.

E10. R. R. Kureshi, B. K. Mishra, D. Thakker, R. John, A. Walker, S. Simpson, N. Thakkar and A. K. Wante, "Data-Driven Techniques for Low-Cost Sensor Selection and Calibration for the Use Case of Air Quality Monitoring," *Sensors*, vol. 22, no. 3, p. 1093, 2022.

E11. Q. Lu, Y. Xu, Z. Ni, C. Wu and H. Huo, "Optimizing the Calibration Parameter of an Infrared Hyperspectral Interferometer Using Simulated Data," *IEEE Geoscience and Remote Sensing Letters*, vol. 19, pp. 1–5, 2022.

E12. J. Borges, F. Bastos, I. Correa, P. Batista and A. Klautau, "CAVIAR: Co-Simulation of 6G Communications, 3-D Scenarios, and AI for Digital Twins," *IEEE Internet of Things Journal*, vol. 11, no. 19, pp. 31287–31300, 1 Oct. 2024.

E13. S. Kabir, C. Wagner and Z. Ellerby, "Toward Handling Uncertainty-At-Source in AI—A Review and Next Steps for Interval Regression," *IEEE Transactions on Artificial Intelligence*, vol. 5, no. 1, pp. 3–22, Jan. 2024.

E14. Z. Xie, F. He, S. Fu, I. Sato, D. Tao and M. Sugiyama, "Artificial Neural Variability for Deep Learning: On Overfitting, Noise Memorization, and Catastrophic Forgetting," *Neural Computation*, vol. 33, no. 8, pp. 2163–2192, 26 July 2021.

E15. Z. Chen et al., "Data-Driven Methods Applied to Soft Robot Modeling and Control: A Review," *IEEE Transactions on Automation Science and Engineering*, vol. 22, pp. 2241–2256, 2025.

E16. S. A. I. Alfarozi, K. Pasupa, M. Sugimoto and K. Woraratpanya, "Local Sigmoid Method: Non-Iterative Deterministic Learning Algorithm for Automatic Model Construction of Neural Network," *IEEE Access*, vol. 8, pp. 20342–20362, 2020.

E17. K. Kayabol, "Approximate Sparse Multinomial Logistic Regression for Classification," *IEEE Transactions on Pattern Analysis and Machine Intelligence*, vol. 42, no. 2, pp. 490–493, Feb. 2020.

E18. R. Wu, S. D. Hamshaw, L. Yang, D. W. Kincaid, R. Etheridge and A. Ghasemkhani, "Data Imputation for Multivariate Time Series Sensor Data with Large Gaps of Missing Data," *IEEE Sensors Journal*, vol. 22, no. 11, pp. 10671–10683, June 2022.

E19. F. A. Castillo, J. D. Sweeney and W. E. Zirk, "Using Evolutionary Algorithms to Suggest Variable Transformations in Linear Model Lack-of-Fit Situations," *Proceedings of the 2004 Congress on Evolutionary Computation*, Portland, OR, USA, vol. 1, pp. 556–560, 2004.

E20. M. I. Habibie and N. Nurda, "Performance Analysis and Classification Using Naïve Bayes and Logistic Regression on Big Data," *2022 1st International Conference on Smart Technology, Applied Informatics, and Engineering (APICS)*, Surakarta, Indonesia, pp. 48–52, 2022.

E21. S. Bobek, M. Kuk, M. Szelążek and G. J. Nalepa, "Enhancing Cluster Analysis with Explainable AI and Multidimensional Cluster Prototypes," *IEEE Access*, vol. 10, pp. 101556–101574, 2022.

E22. K. P. Sinaga and M. -S. Yang, "Unsupervised K-Means Clustering Algorithm," *IEEE Access*, vol. 8, pp. 80716–80727, 2020.

E23. S. Ueno and O. Sakai, "Data Driven Calibration of Color-Sensitive Optical Sensor by Supervised Learning for Botanical Application," *IEEE Sensors Journal*, vol. 22, no. 12, pp. 11915–11927, June 2022.

E24. S. Tondini, R. Scilla and P. Casari, "Minimized Training of Machine Learning-Based Calibration Methods for Low-Cost O3 Sensors," *IEEE Sensors Journal*, vol. 24, no. 3, pp. 3973–3987, Feb. 2024.

E25. S. K. Moore, D. Schneider and E. Strictland, "How Deep Learning Works: Inside the Neural Networks That Power Today's AI," *IEEE Spectrum*, vol. 58, no. 10, pp. 32–33, 2021.

E26. H. Eren, C. C. Fung, K. W. Wong and A. Gupta, "Artificial Neural Networks in Estimation of Hydrocyclone Parameter d50c with Unusual Input Variables," *IEEE Transactions on Instrumentation and Measurement*, vol. 46, no. 4, pp. 908–912, 1997.

E27. S. S. Mittal, "Integration of Human Knowledge with AI through NN," *2023 3rd International Conference on Advanced Computing and Innovative Technologies in Engineering (ICACITE)*, Greater Noida, India, pp. 2272–2276, 2023.

E28. J. R. M. Soares De Souza, A. De Conti and C.L. De Castro, "Using Reduction Techniques to Obtain a Minimal Consistent Data Set for Training Artificial Neural Networks for Fault Identification Problems," *2023 Workshop on Communication Networks and Power Systems (WCNPS)*, Brasilia, Brazil, pp. 1–7, 2023.

E29. T. Maresa, E. Janouchova and A. Kucerova, "Artificial Neural Networks in the Calibration of Nonlinear Mechanical Models," *Advances in Engineering Software*, vol. 95, pp. 68–81, May 2016.

E30. C. Liu, C. Zhao, Y. Wang and H. Wang," Machine Learning Based Calibration of Temperature Sensors," *Sensors*, vol. 23, no. 7347, pp. 1–13, 2023.

E31. M. Ragab, E. Eldele, Z. Chen, M. Wu, C. -K. Kwoh and X. Li, "Self-Supervised Autoregressive Domain Adaptation for Time Series Data," *IEEE Transactions on Neural Networks and Learning Systems*, vol. 35, no. 1, pp. 1341–1351, Jan. 2024.

E32. M. Ntemi, I. Sarridis and C. Kotropoulos, "An Autoregressive Graph Convolutional Long Short-Term Memory Hybrid Neural Network for Accurate Prediction of COVID-19 Cases," *IEEE Transactions on Computational Social Systems*, vol. 10, no. 2, pp. 724–735, April 2023.

E33. K. M. Cohen, S. Park, O. Simeone and S. Shamai Shitz, "Calibrating AI Models for Wireless Communications via Conformal Prediction," *IEEE Transactions on Machine Learning in Communications and Networking*, vol. 1, pp. 296–312, 2023.

E34. Y. Fukuchi and S. Yamada, "Dynamic Selection of Reliance Calibration Cues with AI Reliance Model," *IEEE Access*, vol. 11, pp. 138870–138881, 2023.

E35. K. Y. W. Scheper and G. C. H. E. de Croon, "Abstraction, Sensory-Motor Coordination, and the Reality Gap in Evolutionary Robotics," *Artificial Life*, vol. 23, no. 2, pp. 124–141, May 2017.

E36. H. Mostafa, "Supervised Learning Based on Temporal Coding in Spiking Neural Networks," *IEEE Transactions on Neural Networks and Learning Systems*, vol. 29, no. 7, pp. 3227–3235, July 2018.

E37. I. -M. Comşa, K. Potempa, L. Versari, T. Fischbacher, A. Gesmundo and J. Alakuijala, "Temporal Coding in Spiking Neural Networks with Alpha Synaptic Function: Learning with Backpropagation," *IEEE Transactions on Neural Networks and Learning Systems,* vol. 33, no. 10, pp. 5939–5952, Oct. 2022.

E38. P. Katti, N. Skatchkovsky, O. Simeone, B. Rajendran and B. M. Al-Hashimi, "Bayesian Inference on Binary Spiking Networks Leveraging Nanoscale Device Stochasticity," *2023 IEEE International Symposium on Circuits and Systems (ISCAS),* Monterey, CA, USA, pp. 1–5, 2023.

E39. T. M. Hossain, M. Hermana and S. J. Abdulkadir, "Epistemic Uncertainty and Model Transparency in Rock Facies Classification Using Monte Carlo Dropout Deep Learning," *IEEE Access,* vol. 11, pp. 89349–89358, 2023.

E40. D. B. Wang, L. Feng and M. L. Zhang, "Rethinking Calibration of Deep Neural Networks: Do Not Be Afraid of Overconfidence," *35th Conference on Neural Information Processing Systems (NeurIPS 2021),* Online Conference, Canada, pp. 1–12, 2021.

E41. J. -J. Liu, Q. Hou, M. -M. Cheng, C. Wang and J. Feng, "Improving Convolutional Networks with Self-Calibrated Convolutions," *2020 IEEE/CVF Conference on Computer Vision and Pattern Recognition (CVPR),* Seattle, WA, USA, pp. 10093–10102, 2020.

E42. Z. Wang, C. M. Wong, B. Wang, Z. Feng, F. Cong and F. Wan, "Compact Artificial Neural Network Based on Task Attention for Individual SSVEP Recognition with Less Calibration," *IEEE Transactions on Neural Systems and Rehabilitation Engineering,* vol. 31, pp. 2525–2534, 2023.

E43. C. C. Chen et al., "Calibration of Low-Cost Particle Sensors by Using Machine-Learning Method," *2018 IEEE Asia Pacific Conference on Circuits and Systems (APCCAS),* Chengdu, China, pp. 111–114, 2018.

E44. M. Ö. Efe, B. Kürkçü, C. Kasnakoğlu, Z. Mohamed and Z. Liu, "A Modified Levenberg Marquardt Algorithm for Simultaneous Learning of Multiple Datasets," *IEEE Transactions on Circuits and Systems II: Express Briefs,* vol. 71, no. 4, pp. 2379–2383, April 2024.

E45. J. Liu and D. Zhou, "Minimum Functional Length Analysis of K-Mer Based on BPNN," *IEEE/ACM Transactions on Computational Biology and Bioinformatics,* vol. 19, no. 5, pp. 2920–2925, Sept.–Oct. 2022.

E46. N. B. Gaikwad, V. Tiwari, A. Keskar and N. C. Shivaprakash, "Efficient FPGA Implementation of Multilayer Perceptron for Real-Time Human Activity Classification," *IEEE Access,* vol. 7, pp. 26696–26706, 2019.

E47. N. Nguyen and K. -C. Chen, "Bayesian Quantum Neural Networks," *IEEE Access,* vol. 10, pp. 54110–54122, 2022.

E48. M. Joshaghani, A. Davari, F. N. Hatamian, A. Maier and C. Riess, "Bayesian Convolutional Neural Networks for Limited Data Hyperspectral Remote Sensing Image Classification," *IEEE Geoscience and Remote Sensing Letters,* vol. 20, pp. 1–5, Article no. 5506305, 2023.

E49. P. R. Genssler, A. Vas and H. Amrouch, "Brain-Inspired Hyperdimensional Computing: How Thermal-Friendly for Edge Computing?" *IEEE Embedded Systems Letters,* vol. 15, no. 1, pp. 29–32, March 2023.

E50. H. Eren, *Artificial Intelligence in Wireless Sensors and Instruments: Networks and Applications,* CRC Press, Boca Raton, USA, 2025.

E51. J. Jiao, X. Sun, L. Fang and J. Lyu, "An Overview of Wireless Communication Technology Using Deep Learning," *China Communications,* vol. 18, no. 12, pp. 1–36, Dec. 2021.

E52. Z. Li, F, Liu, W. Yang, S. Peng and J. Zhou, "A Survey of Convolution Neural Networks: Analysis, Applications, and Prospects" *IEEE Transactions on Neural Networks and Learning Systems,* vol. 33, no. 12, pp. 6999–7019, Dec. 2022.

E53. Y. Li, Z. Yao, L. Mao, B. Shi, L. Yao and J. Song, "CNN-Assisted Adaptive Signal Decomposition Method for Correcting the Distorted Calibration Signals of Pressure Sensors," *IEEE Transactions on Instrumentation and Measurement*, vol. 73, pp. 1–10, 2024.

E54. M. Devanathan, D. Prasannan, P. Singh, O. Joshua, H. Adity Pai and P. Jakhar, "Effective Techniques Non-Linear Dynamic Model Calibration Using CNN," *2024 IEEE International Conference on Computing, Power and Communication Technologies (IC2PCT)*, Greater Noida, India, pp. 832–836, 2024.

E55. A. M. Seba, K. A. Gemeda and P. J. Ramulu, "Prediction and Classification of IoT Sensor Faults Using Hybrid Deep Learning Model," *Discover Applied Sciences*, vol. 6, no. 1, p. 9, 2024.

E56. H. Zhou, W. Chen, L. Cheng, J. Liu and M. Xia, "Trustworthy Fault Diagnosis with Uncertainty Estimation through Evidential Convolutional Neural Networks," *IEEE Transactions on Industrial Informatics*, vol. 19, no. 11, pp. 10842–10852, Nov. 2023.

E57. A. Lanning, A. E. Zaghi and T. Zhang, "Applicability of Convolutional Neural Networks for Calibration of Nonlinear Dynamic Models of Structures," *Frontiers in Built Environment*, vol. 8, Article 873546, p. 16, April 2022.

E58. H. Su et al., "Towards Model-Free Tool Dynamic Identification and Calibration Using Multi-Layer Neural Network," *Sensors*, vol. 19, no. 17, p. 3636, 2019.

E59. N. Mustaqeem and S. Kwon "CNN-Assisted Enhanced Audio Signal Processing for Speech Emotion Recognition," *Sensors*, vol. 183, pp. 1–15, 2020.

E60. K. Balaskas et al., "Hardware-Aware DNN Compression via Diverse Pruning and Mixed-Precision Quantization," *IEEE Transactions on Emerging Topics in Computing*, vol. 12, no. 4, pp. 1079–1092, Oct.–Dec. 2024.

E61. J. H. Ahn, Y. Ma, S. Park and C. You, "Federated Active Learning (F-AL): An Efficient Annotation Strategy for Federated Learning" *IEEE Access*, vol. 12, pp. 39261–39269, 2024.

E62. A. Payandeh, K. T. Baghaei, P. Fayyazsanavi, S. B. Ramezani, Z. Chen and S. Rahimi, "Deep Representation Learning: Fundamentals, Technologies, Applications, and Open Challenges," *IEEE Access*, vol. 11, pp. 137621–137659, 2023.

E63. M. Khannefer and S. Shirmohammadi, "Applied AI in Instrumentation and Measurement: The Deep Learning Revolution," *IEEE Instrumentation and Measurement Magazine*, vol. 23, no. 6, pp. 10–17, Sept. 2020.

E64. A. M. Alam, M. Kurum, M. Ogut and A. C. Gurbuz, "Microwave Radiometer Calibration Using Deep Learning with Reduced Reference Information and 2-D Spectral Features," *IEEE Journal of Selected Topics in Applied Earth Observations and Remote Sensing*, vol. 17, pp. 748–765, 2024.

E65. S. Wu, A. Hadachi, D. Vivet and Y. Prabhakar, "NetCalib: A Novel Approach for LiDAR-Camera Auto-Calibration Based on Deep Learning," *2020 25th International Conference on Pattern Recognition (ICPR)*, Milan, Italy, pp. 6648–6655, 2021.

E66. T. Xiang et al., "Federated Learning with Dynamic Epoch Adjustment and Collaborative Training in Mobile Edge Computing," *IEEE Transactions on Mobile Computing*, vol. 23, no. 5, pp. 4092–4106, May 2024.

E67. J. Chen and X. Ran, "Deep Learning with Edge Computing: A Review," *Proceedings of the IEEE*, vol. 107, no. 8, pp. 1655–1674, Aug. 2019.

E68. S. Wu, A. Hadachi, D. Vivet and Y. Prabhakar, "This Is the Way: Sensors Auto-Calibration Approach Based on Deep Learning for Self-Driving Cars," *IEEE Sensors Journal*, vol. 21, no. 24, pp. 27779–27788, Dec. 2021.

E69. M. Ogut et al., "Deep Learning Approach for Microwave and Millimeter-Wave Radiometer Calibration," *IEEE Transactions on Geoscience and Remote Sensing*, vol. 57, no. 8, pp. 5344–5355, March 2019.

E70. X. Wang et al., "Deep Reinforcement Learning: A Survey," *IEEE Transaction on Neural Networks and Learning Systems,* vol. 35, no. 4, pp. 5064–5078, April 2024.

E71. P. Suyal, S. Dutt, R. Sharma and L. Mohan, "An Agile Review of Machine Learning Technique," *2022 11th International Conference on System Modelling and Advancement in Research Trends (SMART),* Moradabat, India, pp. 75–79, 2022.

E72. J. Wang, A. Pal, Q. Yang, K. Kant, K. Zhu and S. Guo, "Collaborative Machine Learning Schemes, Robustness, and Privacy" *IEEE Transactions on Neural Networks and Machine Learning,* vol. 34, no. 12, pp. 9625–9642, Dec. 2023.

E73. B. Yang, C. Peng and F. Jiang, "Model Calibration of Active Magnetic Bearing Using Deep Reinforcement Learning," *2023 7th International Symposium on Computer Science and Intelligent Control (ISCSIC),* Nanjing, China, pp. 16–20, 2023.

E74. Y. Tian, M. A. Chao, C. Kulkarni, K. Goebel and O. Fink, "Real-Time Model Calibration with Deep Reinforcement Learning," *Mechanical Systems and Signal Processing,* vol. 165, pp. 108284, 2022.

E75. K. Krishna, S. L. Brunton and Z. Song, "Finite Time Lyapunov Exponent Analysis of Model Predictive Control and Reinforcement Learning," *IEEE Access,* vol. 11, pp. 118916–118930, 2023.

E76. J. Ren et al., "High Precision Calibration Algorithm for Binocular Stereo Vision Camera Using Deep Reinforcement Learning," vol. 2022, Article ID 6596868, 10 pages, 2022.

E77. A. Ashiquzzaman, H. Lee, T. -W. Um and J. Kim, "Energy-Efficient IoT Sensor Calibration with Deep Reinforcement Learning," *IEEE Access,* vol. 8, pp. 97045–97055, 2020.

E78. T. Liu, B. Tian, Y. Ai and F. -Y. Wang, "Parallel Reinforcement Learning-Based Energy Efficiency Improvement for a Cyber-Physical System," *IEEE/CAA Journal of Automatica Sinica,* vol. 7, no. 2, pp. 617–626, March 2020.

E79. T. H. L. Nguyen and S. Park, "Intelligent Ultrasonic Flow Measurement Using Linear Array Transducer with Recurrent Neural Networks," *IEEE Access,* vol. 8, pp. 137564–137573, 2020.

E80. N. Laoué, A. Lepers, L. Deletraz and C. Faure, "Neural Network Calibration of Airborne Magnetometers," *2023 IEEE 10th International Workshop on Metrology for AeroSpace (MetroAeroSpace),* Milan, Italy, pp. 37–42, 2023.

E81. Y. Shi and Z. Rong, "Anaysis of Q-Learning Like Algorithms through Evoltionay Game Dynamics," *IEEE Transactions on Circuits and Systems II: Express Beliefs,* vol. 69, no. 5, pp. 2463–2467, May 2022.

E82. M. N. H. Nguyen et al., "Self-Organizing Democratized Learning: Toward Large Scale Distributed Learning Systems," *IEEE Transactions on Neural Networks and Machine Learning,* vol. 34, no. 12, pp. 10698–10710, Dec. 2023.

E83. Z. Cai, J. Chen, Y. Fan, Z. Zheng and K. Li, "Blockchain-Empowered Federated Learning: Benefits, Challenges, and Solutions," ArXiv abs/2403.00873, 2024.

E84. X. Yang, B. Xiong, Y. Huang and C. Xu, "Cross-Modal Federated Human Activity Recognition," *IEEE Transactions on Pattern Analysis and Machine Intelligence,* vol. 46, no. 8, pp. 5345–5361, Aug. 2024.

E85. Q. Wang, W. Liao, Y. Guo, M. McGuire and W. Yu, "Blockchain-Empowered Federated Learning through Model and Feature Calibration," *IEEE Internet of Things Journal,* vol. 11, no. 4, pp. 5770–5780, Feb. 2024.

E86. C. Li et al., "Federated Transfer Learning for On-Device LLMs Efficient Fine-Tuning Optimization," *Big Data Mining and Analytics,* vol. 8, no. 2, pp. 430–446, April 2025.

E87. T. Castiglia, S. Wang and S. Patterson, "Flexible Vertical Federated Learning with Heterogeneous Parties," *IEEE Transactions on Neural Networks and Learning Systems,* vol. 35, no. 12, pp. 17878–17892, Dec. 2024.

E88. Sumitra, J. Sharma and M. V. Shenoy, "HAFedL: A Hessian-Aware Adaptive Privacy Preserving Horizontal Federated Learning Scheme for IoT Applications," *IEEE Access*, vol. 12, pp. 126738–126753, 2024.

E89. N. Gothoskar et al., "DURableVS: Data-Efficient Unsupervised Recalibrating Visual Servoing via Online Learning in a Structured Generative Model," *2022 International Conference on Robotics and Automation (ICRA)*, Philadelphia, PA, USA, pp. 6674–6680, 2022.

E90. I. Hong and J. Ha, "Generative Adversarial Networks for Solving Hand-Eye Calibration without Data Correspondence," *IEEE Robotics and Automation Letters*, vol. 10, no. 3, pp. 2494–2501, March 2025.

E91. Z. Pan, W. Yu, X. Yi. A. Khan, F. Yuan and Y. Zheng, "Recent Progress on Generative Adversarial Networks (GANs): A Survey," *IEEE Access*, vol. 7, pp. 36322–36333, 2019.

E92. C. Park, Y. Kim, J. -G. Park, D. Hong and C. Seo, "Evaluating Differentially Private Generative Adversarial Networks over Membership Inference Attack," *IEEE Access*, vol. 9, pp. 167412–167425, 2021.

E93. P. Sawangjai et al., "EEGANet: Removal of Ocular Artifacts from the EEG Signal Using Generative Adversarial Networks," *IEEE Journal of Biomedical and Health Informatics*, vol. 26, no. 10, pp. 4913–4924, Oct. 2022.

E94. S. Li, K. Zhang, Q. Cheng, S. Wang and S. Zhang, "Feature Selection for High Dimensional Data Using Weighted K-Nearest Neighbors and Genetic Algorithms," *IEEE Access*, vol 8, pp. 139512–129528, 2020.

E95. Y. Hamed et al., "Two Step Hybrid Calibration Algorithm of Support Vector Regression and K Nearest Neighbors," *Alexandria Engineering Journal*, vol. 59, no. 3, pp. 1181–1190, June 2020.

E96. B. Mahesh, "Machine Learning Algorithms – A Review," *International Journal of Science and Research (IJSR)*, vol. 9, no. 1, pp. 381–386, Jan. 2020.

E97. T. Young, D. Hazarika, S. Poria and E. Cambria, "Recent Trends in Deep Learning-Based Natural Language Processing," *IEEE Computational Intelligence Magazine*, vol. 13, no. 3, pp. 55–75, Aug. 2018.

E98. P. Danenas and T. Skersys, "Exploring Natural Language Processing in Model-to-Model Transformations," *IEEE Access*, vol. 10, pp. 116942–116958, 2022.

E99. Y. Chen et al., "NetGPT: An AI-Native Network Architecture for Provisioning Beyond Personalized Generative Services," *IEEE Network*, vol. 38, no. 6, pp. 404–413, Nov. 2024.

E100. B. Sayin et al., "Crowd-Powered Hybrid Classification Services: Calibration is All You Need," *2021 IEEE International Conference on Web Services (ICWS)*, Chicago, IL, USA, 2021.

E101. A. Kumar, A. Perrault and D. S. Williamson, "Using RLHF to Align Speech Enhancement Approaches to Mean-Opinion Quality Scores," *ICASSP 2025 – 2025 IEEE International Conference on Acoustics, Speech and Signal Processing (ICASSP)*, Hyderabad, India, 2025.

E102. C. Puerto-Santana, P. Larranaga and C. Bileza, "Autoregressive Asymmetric Linear Gaussian Hidden Markov Models," *IEEE Transactions on Pattern Analysis and Machine Intelligence*, vol. 44, no. 9, pp. 4642–4658, Sept. 2022.

E103. N. Yang, J. Xiong, C. Guo, S. Guo and G. Li, "Reflection Coefficients Inversion Based on the Bidirectional Long Short-Term Memory Network," *IEEE Geoscience and Remote Sensing Letters*, vol. 19, pp. 1–5, 2022.

E104. R. M. Samant, M. R. Bachute, S. Gite and K. Kotecha, "Framework for Deep Learning-Based Language Models Using Multi-Task Learning in Natural Language Understanding: A Systematic Literature Review and Future Directions," *IEEE Access*, vol. 10, pp. 17078–17097, 2022.

E105. A. Bhaskar, "Multi-Heuristic State Space Search," *2018 9th Annual Information Technology, Electronics and Mobile Communication Conference (IEMCON)*, Vancouver, BC, Canada, pp. 1251–1255, 2018.

E106. Y. Ozaki, S. Takenaga and M. Onishi, "Global Search versus Local Search in Hyperparameter Optimization," *2022 IEEE Congress on Evolutionary Computing*, Padua, Italy, pp. 1–9, 2022.

E107. A. Das, M. Roopaei, M. Jamshidi and P. Najafirad, "Distributed AI-Driven Search Engine on Visual Internet-of-Things for Event Discovery in the Cloud," *2022 IEEE 17th Annual Systems of Systems Engineering Conference (SOSE)*, Rochester, NY, pp. 514–521, 2022.

E108. H. Mohammadi, M. Razaviyayn and M. R. Jovanović, "Robustness of Accelerated First-Order Algorithms for Strongly Convex Optimization Problems," *IEEE Transactions on Automatic Control*, vol. 66, no. 6, pp. 2480–2495, June 2021.

E109. B. Alzalg and H. Alioui, "Applications of Stochastic Mixed-Integer Second-Order Cone Optimization," *IEEE Access*, vol. 10, pp. 3522–3547, 2022.

E110. O. N. Oyelade, A. E. -S. Ezugwu, T. I. A. Mohamed and L. Abualigah, "Ebola Optimization Search Algorithm: A New Nature-Inspired Metaheuristic Optimization Algorithm," *IEEE Access*, vol. 10, pp. 16150–16177, 2022.

E111. D. F. Resende, L. R. M. Silva, E. G. Nepomuceno and C. A. Duque, "Optimizing Instrument Transformer Performance through Adaptive Blind Equalization and Genetic Algorithms," *Energies*, vol. 16, p. 7354, 2023.

E112. L. Avanthey and L. Beaudoin, "Underwater Calibration in Near Real Time: Focus on Detection Optimized by AI and Selection of Calibration Patterns," *IGARSS 2020 – 2020 IEEE International Geoscience and Remote Sensing Symposium*, Waikoloa, HI, USA, pp. 1576–1579, 2020.

E113. B. Kim and M. Shin, "A Novel Neural-Network Device Modeling Based on Physics-Informed Machine Learning," *IEEE Transactions on Electron Devices*, vol. 70, no. 11, pp. 6021–6025, Nov. 2023.

E114. J. Li and S. X. Yang, "A Novel Feature Learning-Based Bio-Inspired Neural Network for Real-Time Collision-Free Rescue of Multirobot Systems," *IEEE Transactions on Industrial Electronics*, vol. 71, no.11, pp.14420–14429, Nov. 2024.

E115. T. Shen, C. S. Mishra, J. Sampson, M. T. Kandemir and V. Narayanan, "An Efficient Edge-Cloud Partitioning of Random Forest for Distributed Sensor Networks," *IEEE Embedded Systems Letters*, vol. 16, no. 1, pp. 21–24, March 2024.

E116. A. Walia, A. Paliwal, S. Patidar and R. Mahto, "Prediction of Air Quality Index Using Random Forest and Prophet Tool," *2024 19th Annual System of Systems Engineering Conference (SoSE)*, Tacoma, WA, USA, 2024.

E117. K. M. O. Vale, A. C. Gorgônio, F. D. L. E. Gorgônio and A. M. D. P. Canuto, "An Efficient Approach to Select Instances in Self-Training and Co-Training Semi-Supervised Methods," *IEEE Access*, vol. 10, pp. 7254–7276, 2022.

E118. M. -C. Xu et al., "MisMatch: Calibrated Segmentation via Consistency on Differential Morphological Feature Perturbations with Limited Labels," *IEEE Transactions on Medical Imaging*, vol. 42, no. 10, pp. 2988–2999, Oct. 2023.

E119. Y. Chen et al., "BEVSOC: Self-Supervised Contrastive Learning for Calibration-Free BEV 3-D Object Detection," *IEEE Internet of Things Journal*, vol. 11, no. 12, pp. 22167–22182, June, 2024.

E120. A. Tendle, A. Little, S. Scott and M. R. Hasan, "Self-Supervised Learning in the Twilight of Noisy Real-World Datasets," *2022 21st IEEE International Conference on Machine Learning and Applications (ICMLA)*, Nassau, Bahamas, pp. 461–464, 2022.

E121. M. Praveena and V. Jaiganesh, "Literature Review on Supervised Machine Learning Algorithms and Boosting Process," *International Journal of Computer Applications*, vol. 169, no. 8, pp. 32–35, 2017.

E122. T. Furuya and R. Ohbuchi, "DeepDiffusion: Unsupervised Learning of Retrieval-Adapted Representations via Diffusion-Based Ranking on Latent Feature Manifold," *IEEE Access*, vol. 10, pp. 116287–116301, 2022.

E123. D. Onita, "Active Learning Based on Transfer Learning Techniques for Text Classification," *IEEE Access*, vol. 11, pp. 28751–28761, 2023.

E124. Z. Chen et al., "Survey on AI Sustainability: Emerging Trends on Learning Algorithms and Research Challenges," *IEEE Computational Intelligence Magazine*, vol. 18, no. 2, pp. 60–77, May 2023.

E125. W. Yang, S. Li, Z. Li and X. Luo, "Highly Accurate Manipulator Calibration via Extended Kalman Filter-Incorporated Residual Neural Network," *IEEE Transactions on Industrial Informatics*, vol. 19, no. 11, pp. 10831–10841, Nov. 2023.

E126. D. A. Cucci, L. Voirol, M. Khaghani and S. Guerrier, "On Performance Evaluation of Inertial Navigation Systems: The Case of Stochastic Calibration," *IEEE Transactions on Instrumentation and Measurement*, vol. 72, pp. 1–17, 2023.

E127. D. Isla-Cernadas, M. Fernández-Delgado, E. Cernadas, M. S. Sirsat, H. Maarouf and S. Barro, "Closed-Form Gaussian Spread Estimation for Small and Large Support Vector Classification," *IEEE Transactions on Neural Networks and Learning Systems*, vol. 36, no. 3, pp. 4336–4344, March 2025.

E128. C. N. Li, Y. Li and Y. H. Shao, "Large-Scale Structured Output Classification via Multiple Structured Support Vector Machine by Splitting," *IEEE Transaction on Emerging Topics in Computational Intelligence*, vol. 8, no. 2, pp. 2112–2124, April 2024

E129. H. Kuzuno and T. Yamauchi, "Mitigation of Kernel Memory Corruption Using Multiple Kernel Memory Mechanism," *IEEE Access*, vol. 9, pp. 111651–111665, 2021.

E130. Z. Zhu, K. Lin, A. K. Jain and J. Zhou, "Transfer Learning in Deep Reinforcement Learning: A Survey," *IEEE Transactions on Pattern Analysis and Machine Intelligence*, vol. 45, no. 11, pp. 13344–13362, Nov. 2023.

E131. P. P. Ray, "A Survey Cloud Platforms," *Future Computing and Informatics Journal*, vol. 1, no. 1–2, pp. 35–46, 2018.

E132. A. Al-Ridhawi, S. Otoum, M. Aloqaily and A. Boukerche, "Generalizing AI: Challenges and Opportunities for Plug and Play AI Solutions," *IEEE Network*, vol. 35, no. 1, pp. 372–379, Jan. 2021.

E133. A. E. Bouaouad, A. Cherradi, S. Assoul and N. Souissi, "Architectures and Emerging Trends in Internet of Things and Cloud Computing: A Literature Review," *2020 4th International Conference on Advanced Systems and Emergent Techologies (IC_ASET)*, Hammamet, Tunisia, pp. 147–151, Dec. 2021.

E134. S. Nastic, P. Raith, A. Furutanpey, T. Pusztai and S. Dustdar, "A Serverless Computing Fabric for Edge and Cloud," *2022 Proceedings of IEEE 4th International Conference on Cognitive Machine Intelligence (CogMI)*, Virtual Conference, pp. 1–12, 2022.

E135. V. Kjorveziroski, S. Filiposka and V. Trajkocic, "IoT Serverless Computing at Edge: Open Issues and Research Directions," *Computers*, vol. 10, no. 10, Article no. 130, p. 1–21, 2021.

E136. P. Bellavista, J. Berrocal, A. Corradi, K. Sajal and A. Zanni, "A Survey Fog Computing for the Internet of Things," *Journal of Pervasive and Mobile Computing*, vol. 52, pp. 71–99, 2018.

E137. J. R. McClean, "From Molecules to Quantum Computers: A Research Retrospective," *Computing in Science and Engineering*, vol. 23, no. 6, pp. 52–57, Dec. 2021.

E138. H. Y. Huang et al., "Power of Data in Quantum Machines," *Nature Communications*, vol. 12, no. 1, pp. 1–6, 2021.

E139. J. R. McClean, S. Boixo, V. N. Smelyanskiy, R. Babbush and H. Neven, "Barren Plateaus in Quantum Neural Network Training Landscape," *Nature Communications*, vol. 9, no. 1, pp. 1–6, 2018.

E140. P. C. Wu et al., "An Integer-Floating-Point Dual-Mode Gain-Cell Computing-in-Memory Macro for Advanced AI Edge Chips," *IEEE Journal of Solid-State Circuits*, vol. 60, no. 1, pp. 158–170, Jan. 2025.

E141. M. Zhang, H. Zhang, C. Zhang and D. Yuan, "Communication Efficient Deep Compressed Sensing for Edge-Cloud Collaborative Industrial IoT Networks," *IEEE Transactions on Industrial Informatics*, vol. 19, no. 5, pp. 6613–6623, May 2023.

E142. Y. He and L. Xiao, "Structured Pruning for Deep Convolution Neural Networks: A Survey," *IEEE Transactions on Pattern Analysis and Machine Intelligence*, vol. 46, no. 5, pp. 2900–2919, May 2024.

E143. Z. Luo et al., "Improving Data Analytics with Fast and Adaptive Regularization," *IEEE Transactions on Knowledge and Data Engineering*, vol. 33, no. 2, pp. 551–568, Feb. 2021.

E144. R. Novkin, F. Klemme and H. Amrouch, "Approximate- and Quantization Aware Training for Graph Neural Networks," *IEEE Transactions on Computers*, vol. 73, no. 2, pp. 599–612, Feb. 2024.

E145. T. K. S. Flores, M. Medeiros, M. Silva, D. G. Costa and I. Silva, "Enhanced Vector Quantization for Embedded Machine Learning: A Post-Training Approach with Incremental Clustering," *IEEE Access*, vol. 13, pp. 17440–17456, 2025.

E146. H. J. Park, W. Shin, J. S. Kim and S. W. Han, "Leveraging Non-Causal Knowledge via Cross-Network Knowledge Distillation for Real-Time Speech Enhancement," *IEEE Signal Processing Letters*, vol. 31, pp. 1129–1133, 2024.

E147. E. S. Jeon, M. P. Buman and P. Turaga, "Uncertainty-Aware Topological Persistence Guided Knowledge Distillation on Wearable Sensor Data," *IEEE Internet of Things Journal*, vol. 11, no. 18, pp. 30413–30429, Sept. 2024.

E148. C. Li, X. Wang, W. Dong, J. Yan, Q. Liu and H. Zha, "Joint Active Learning with Feature Selection via CUR Matrix Decomposition," *IEEE Transactions on Pattern Analysis and Machine Intelligence*, vol. 41, no. 6, pp. 1382–1396, June 2019.

E149. X. Hu, Y. Han and Z. Geng, "A Novel Matrix Completion Model Based on the Multi-Layer Perceptron Integrating Kernel Regularization," *IEEE Access*, vol. 9, pp. 67042–67050, 2021.

E150. A. Gormez and E. Koyuncu, "Class Based Thresholding in Early Exit Semantic Segmentation Networks," *IEEE Signal Processing Letters*, vol. 31, pp. 1184–1188, 2024.

E151. E. Baccarelli, M. Scarpiniti, A. Momenzadeh and S. S. Ahrabi, "Learning-in-the-Fog (LiFo): Deep Learning Meets Fog Computing for the Minimum-Energy Distributed Early-Exit of Inference in Delay-Critical IoT Realms," *IEEE Access*, vol. 9, pp. 25716–25757, 2021.

E152. A. V. S. Neto, J. B. Camargo, J. R. Armeida and P. S. Cugnasca, "Safety Assurance of Artificial Intelligence-Based Systems: A Systematic Literature Review on State of Art Guidelines for Future Work," *IEEE Access*, vol. 10, pp. 130733–130770, 2022.

E153. H. Tang, O. Catak, M. Kuzlu, E. Catak and Y. Zhao, "Defending AI-Based Automatic Modulation Recognition Models against Adversarial Attacks," *IEEE Access*, vol. 11, pp. 76629–76637, 2023.

E154. M. Aristodemou, X. Liu, S. Lambotharan and B. AsSadhan, "Bayesian Optimization-Drive Adversarial Poisoning Attacks against Distributed Learning," *IEEE Access*, vol. 11, pp. 86214–86226, 2023.

E155. M. Shateri, F. Messina, F. Labeau and P. Piantanida, "Preserving Privacy in GANs against Membership Inference Attack," *IEEE Transactions on Information Forensics and Security*, vol. 19, pp. 1728–1743, 2024.

E156. I. Erkek and E. Irmak, "Enhancing Cybersecurity of a Hydroelectric Power Plant through Digital Twin Modeling and Explainable AI," *IEEE Access*, vol. 13, pp. 41887–41908, 2025.

E157. A. Roukounaki, S. Efremidis, J. Soldatos, J. Neises, T. Walloschke and N. Kefalakis, "Scalable and Configurable End-to-End Collection and Analysis of IoT Security Data: Towards End-to-End Security in IoT Systems," *2019 Global IoT Summit (GIoTS)*, Aarhus, Denmark, pp. 1–6, 2019.

E158. I. Singh, E. Sherman, D. M. A. Dut and H. Jain, "Network Intrusion Detection Using GAN and Resnet Optimized with Glowworm Optimization," *2023 5th International Conference on Advances in Computing, Communication Control and Networking (ICAC3N)*, Greater Noida, India, pp. 1348–1354, 2023.

E159. R. Brooks, "A Human in the Loop: AI Won't Surpass Human Intelligence Anytime Soon," *IEEE Spectrum*, vol. 58, no. 10, pp. 48–49, Oct. 2021.

E160. J. VerWey, "The Other Artificial Intelligence Hardware Problem," *IEEE Computers*, vol. 55, no. 1, pp. 34–42, Jan. 2022.

E161. H. Mutaba, "NVIDIA Drive Xavier SOC Detailed-A Marvel of Engineering, Biggest and Most SOC Design to Date with 9 Billion Transistors," *2018 WCCF TECH*, Jan. 2018. Available at: https://wccftech.com/nvidia-drive-xavier-soc-detailed/#:~:text=N VIDIA%20has%20just%20unveiled%20the%20full%20extent%20of,of%20the%20 densest%20packed%20SOC%20design%20to%20date (Accessed on 24 July 2025)

E162. C. Deng, X. Fang, X. Wang and K. Law, "Software Orchestrated and Hardware Accelerated Artificial Intelligence: Towards Low Latency Edge Computing," *IEEE Wireless Communications*, vol. 29, no. 4, pp. 110–117, Aug. 2022.

E163. R. Stewart et al., "Optimizing Hardware Accelerator Neural Networks with Quantization and a Distillation Evolutionary Algorithm," *Electronics*, vol. 10, no. 4, Article no. 396, 2021.

E164. J. M. de la Rosa, "AI-Assisted Sigma-Delta Converters—Application to Cognitive Radio," *IEEE Transactions on Circuits and Systems II: Express Briefs*, vol. 69, no. 6, pp. 2557–2563, June 2022.

E165. M. Talib et al., "A Systematic Literature Review on Hardware Implementation of Artificial Intelligence Algorithms," *The Journal of Supercomputing*, vol. 77, pp. 1897–1938, Feb. 2021.

E166. I. Dadras, G. M. Sarda, N. Laubeuf, D. Bhattacharjee and A. Mallik, "AIMC Modeling and Parameter Tuning for Layer-Wise Optimal Operating Point in DNN Inference," *IEEE Access*, vol. 11, pp. 87189–87199, 2023.

E167. A. Nascita, A. Montieri, G. Aceto, D. Ciuonzo, V. Persico and A. Pescapé, "Improving Performance, Reliability, and Feasibility in Multimodal Multitask Traffic Classification with XAI," *IEEE Transactions on Network and Service Management*, vol. 20, no. 2, pp. 1267–1289, June 2023.

E168. DARPA, *XAI: Explainable Artificial Intelligence*, Sept. 2018. https://www.darpa.mil/ research/programs/explainable-artificial-intelligence (Accessed on 20 April 2025).

E169. F. Chen, J. Zhou, A. Holzinger, K. R. Fleischmann and S. Stumpf, "Artificial Intelligence Ethics and Trust: From Principles to Practice," *IEEE Intelligent Systems*, vol. 38, no. 6, pp. 5–8, Nov.–Dec. 2023.

E170. B. Li, P. Goi et al., "Trustworthy AI: From Principles to Practices," *Association for Computing Machinery Computing Surveys*, vol. 1, no. 1, May 2022.

E171. F. M. R. Junior and C. A. Kamienski, "A Survey on Trustworthiness for the Internet of Things," *IEEE Access*, vol. 9, pp. 42493–42514, 2021.

E172. A. Kuznietsov, B. Gyevnar, C. Wang, S. Peters and S. V. Albrecht, "Explainable AI for Safe and Trustworthy Autonomous Driving: A Systematic Review," *IEEE Transactions on Intelligent Transportation Systems*, vol. 25, no. 12, pp. 19342–19364, Dec. 2024.

E173. D. Petkovic, "It is Not "Accuracy vs. Explainability"—We Need Both for Trustworthy AI Systems," *IEEE Transactions on Technology and Society*, vol. 4, no. 1, pp. 46–53, March 2023.

E174. J. R. Carvalko, "Generative AI, Ingenuity, and Law," *IEEE Transactions on Technology and Society*, vol. 5, no. 2, pp. 169–182, June 2024.

E175. O. Wehbi et al., "Enhancing Mutual Trustworthiness in Federated Learning for Data-Rich Smart Cities," *IEEE Internet of Things Journal*, vol. 12, no. 3, pp. 3105–3117, Feb. 2025.

E176. M. Ebadi Jalal and A. Elmaghraby, "Toward Deep Semi-Supervised Continual Learning: A Unified Survey for Scalable and Adaptive AI," *IEEE Access*, vol. 13, pp. 60903–60929, 2025.

E177. N. H. Chapman, F. Dayoub, W. Browne and C. Lehnert, "Predicting Class Distribution Shift for Reliable Domain Adaptive Object Detection," *IEEE Robotics and Automation Letters*, vol. 8, no. 8, pp. 5084–5091, Aug. 2023.

E178. A. K. Raz et al., "Explainable AI and Robustness-Based Test and Evaluation of Reinforcement Learning," *IEEE Transactions on Aerospace and Electronic Systems*, vol. 60, no. 5, pp. 6110–6123, Oct. 2024.

6 Applications of AI in Calibration

6.1 INTRODUCTION

AI offers new frontiers in the calibration of devices and systems. Calibration of a device or a system is achieved as a part of algorithm whenever AI is applied. It provides calibration of individual devices as well as complete system calibration. In this chapter, some examples of AI-related calibrations are provided. Only a few examples are selected in the realm of endless application areas. All types of AI models are used in calibrations, ranging from activity recognition to sensors and sensor networks distributed all around the world, which may be operating collaboratively to perform certain tasks. Some of the application areas are air pollution monitoring, autonomous vehicles, fire detection, and others. Deep learning methods such as a convolutional neural network (CNN), recurrent neural networks (RNNs), and long short-term memory (LSTMs) and their variations are commonly used in the calibration of large-scale systems.

6.2 AI-RELATED CALIBRATION EXAMPLES

6.2.1 ACTIVITY RECOGNITION

Activity recognition has been investigated in many areas, typically on the behavior of the living and human activities. As an example, driver activity recognition is based on sensors and camera image recognition. Diver's body postures, hand movements, alertness analysis, side distractions, and gazing are some of the examples. Other examples of AI-related activity recognition are in healthcare, patient care, disabled care, human–computer interaction, wearables, and sports. AI calibration in activity recognition involves device calibration, temperature scaling, calibration by input guidance, video-based calibrations, and so on. Data-driven methods are trained on datasets to learn patterns and make predictions about unknown activities. Several AI models are used such as CNNs, random forest, and LSTMs.[F1]

6.2.2 AIR QUALITY

Air qualities are determined by IoT-based sensors, meteorological stations, and satellite data. All these methods generate massive amounts of data that are suitable for applications of AI models. Intelligent outdoor air quality sensing and forecasting relies on deployment of a large number of sensors for determining air quality in cities, regions, or geographic areas. Measurements involve many variables such as aerosols, particulate matter, hydrocarbons, soot, CO_2, No_x and other gases. Typical

DOI: 10.1201/9781003590767-6

AI models often used in air quality monitoring are LSTM, CNN, support vector machine (SVM), and transfer learning.[F2) (F3)]

6.2.3 ANTENNAS

Artificial intelligence is used in antenna design and optimization of wireless communications. AI techniques enable the smart antenna target to be learned in an efficient, reliable, and adaptive manner. AI is also used in various antenna optimization, including parametric optimization, reconfigurability, fault detection, and beam steering. Methods include parallel optimization, single and multi-objective optimization, variable fidelity optimization, multilayer ML-assisted optimization, and surrogate-based optimization. AI models enhance antenna behavior prediction, reduce the number of simulations, improve computer efficiency, and speed up the antenna design process. Some AI models applied in antenna are CNN, deep MIMO (massive multiple input multiple output), SVM, random forest, Naïve Bias, and hybrid models. Calibrations are based on ray tracers, electric parameters, focal length, phase correction, power adjustments, antenna arrays, and so on.[F4) (F5)]

6.2.4 AUTONOMOUS VEHICLE

Autonomous driving systems have been developed and are improving almost daily. They find applications in mining, warehouses, passenger vehicles, underwater vehicles, and other road vehicles. AI is the engine behind the wheels, enabling autonomous vehicles to perceive their surroundings, navigate safely, and make decisions in the absence of humans. They are equipped with many multimodal sensors. A typical example is the multiple LiDAR (3D Light Detection and Ranging) to cover the distant and near space of the vehicle. The precision of perception relies on the quality of sensor calibration. Researchers aim to develop an accurate, automatic, and robust calibration strategy for multiple sensory systems. Data-driven methods can generate an "end-to-end" numerical model with a look-up table, which saves the vehicle's velocity, control command, and acceleration. Autonomous vehicles involve a broad range of communication systems and networks. Some AI models used in autonomous vehicles in automation, navigation, sensors, design, and calibration purposes are CNNs, RNNs, generative adversarial networks (GANs), multimodal transfer module, human-in-the-loop, and others.[F6) (F7)]

6.2.5 BLOOD PRESSURE

Artificial intelligence is used in blood pressure (BP) monitoring for risk predictions, treatments, and BP management. Blood pressure is measured using various techniques, cuff methods, camera techniques, wearable devices, electrocardiograms, and noncontact methods such as photoplethysmogram (PPG) signals. PPG sensors can measure the changes in the blood volume directly through contact with the skin. AI models help detect BP during intense exercise, resting, and sleeping. AI methods employed for BP calibration purposes are CNNs, BP-convolutional recurrent NNs, LSTMs, convolution long-short-term Dep NNs, transfer learning, and others.[F8)]

6.2.6 CAMERA CALIBRATION

Camera calibrations are based on different methods. Geometric-based calibrations, self-calibration, and multicamera calibrations are some examples. Self-calibration methods only make use of the corresponding relationship between multiple images, namely, they rely only on the information in an image space. AI camera calibrations eliminate the need for manual adjustments by estimating the intrinsic and extrinsic parameters of images. Camera calibrations are essential in applications such as autonomous vehicles, augmented and virtual reality applications, robotic vision, industrial automation, military, sports, surgery, medical imaging, and traffic control. Some AI-based commercial camera calibration software are Calib, Github, and Android camera calibration.[F9]

6.2.7 CHEMOMETRICS

Chemometrics is a set of mathematical tools and statistical methods used for optimally analyzing multivariate data. The data are generated by experimental designs, simulations or measurement devices. Chemometrics concentrates on noise reduction, interferents, exploratory aspects, and outlier control. Principal component analysis (PCA) is used often in supervised and unsupervised data analysis. Other methods include partial least-squares discriminant analysis, SVM, decision trees, random forest boosting methods, and shallow ANN methods.[F9]

6.2.8 COMPUTATIONAL SENSORS

Computational sensors (simulation of sensors) enabled by machine learning are primarily used in distributed applications that benefit from "big data" analytics and the Internet of Things (IoTs). Computational sensors aim to create sensing networks in various fields, including biomedical diagnostics, environmental sensing, industrial sensor networks, and global health systems. Computational sensors are applied in fluid dynamics, CMOS technologies, wearables, intelligent sensor design, genome analysis, pressure sensing, multimedia sensor networks, optical fibers, RFID, smart grids, and computational geometry. Some of the AI models include CNNs, RNNs, and other deep NNs.[B25]

6.2.9 CYBER-PHYSICAL SYSTEMS

Cyber-physical systems integrated with AI enable intelligent control of systems and autonomous control, such as robots and self-driving vehicles. They are data-driven systems with decision-making capabilities in complex, non-linear, and often multistage environments. Cyber-physical systems can be regarded as the backbone of the fourth industrial revolution, and they find applications in industrial systems, IoTs, building management, robotics, medical devices, manufacturing, and so on. The AI technique evolved as cyber-physical AIs using the diverse range of models, such as GAN-based models, LSTMs, RLs, deep Q-networks, CNNs, RNNs, risk-aware learning, transfer learning, Bayesian learning, and others.[F10]

6.2.10   DEVICE CLUSTERING

Device clustering is a group of similar devices for convenient management, centralized data processing, and control and reliability. Device grouping can be done in two basic ways: hardware clustering and software clustering. Device clustering algorithms are used in tasks such as troubleshooting, calibrations, network management, resource usage, anomaly detection, and so forth. Some device clustering algorithms are K-means clustering, SVM, PCA, density-based spatial clustering of applications with noise, and hierarchical clustering.[F11]

6.2.11   DYNAMIC SYSTEMS

AI methods are used in dynamic operational systems. Dynamic AI models learn and adapt continuously by observing new data and reconfiguring their performances. Calibration-related analyses are made decisions are taken while the system is operating. Typical application areas are biology, economics, power grids, altimeters, fault diagnosis, robotics, and many others. Data-driven models are physically informed machine learning, GANs, and others.[F12]

6.2.12   EARTH OBSERVATION SATELLITES

AI models are used extensively on Earth observation satellites for data processing and calibrations of the devices. Large volumes of data can be efficiently processed on a real-time basis by using AI models. Earth observation Satellites are used in disaster response, environmental monitoring, urban planning, agriculture, forestry, climate change, and pollution monitoring. Satellites are limited by the on-orbital calibration capability, scaling the measured radiance in accuracy and stability. An extensive range of AI algorithms are used in satellite systems, varying from image processing to frequency band identification, some of which are SVMs, KNNs, and gradient boosting machines.[F13]

6.2.13   ECONOMETRICS

Econometrics calibration methods range from probabilistic determination and estimation of model parameters to applications of AI models. Models may be different in dynamic macroeconomics and microeconomics. Many econometric calibration techniques are still evolving, including double calibrations, use of data pre-adjustments, and the incorporation of model estimations. Large datasets used in AI applications lead to finding causal variables and controlling confounding factors. Here, large language models (LLMs) find wider applications in economic forecasting, inflation determination, fine tuning and calibration to determine market trends. Deep reinforcement learning, generative AI, and hybrid models are other typical examples.

6.2.14   EDUCATION

Artificial intelligence is used in all stages of education, from student learning to performance assessments of staff and students. As an example, students and instructors use ChatGPT for grading assignments. Student views of ChatGPT have evolved from a perceived "cheating tool" to a collaborative resource that requires human oversight and calibrated trust. Another example is calibration within higher education, aiming to ensure consistent standards for judging the quality of student work across institutions. AI models in academia are used for adaptive learning, classroom management, tutoring, grading, planning, interactive learning games, proctoring, administration, virtual campus activities, virtual tours, test preps, plagiarism detection, research, and so forth. The use of AI models also indicates attention focus on the related videos, providing valuable insights on deception cues. Some of the AI models used in education are LLMs and all types of AI models.[F14] [F15]

6.2.15   ELECTRICAL ENGINEERING

AI models are used in many areas of electrical engineering, such as power systems, smart grids, renewable energy, distribution network reliability, cybersecurity, R&D, and education and training. Many of these application areas require calibration of devices and systems. For example, maintaining good quality transient stability models for power system planning and operational analysis is essential. Identification and calibration of parameters that work well for multiple events is still a challenging problem. There are many examples of calibrations using AI models in electrical engineering and related areas, some of which are CNNs, LSTMs, DRLs, soft actor critic, and many others.[F16]

6.2.16   ENVIRONMENTAL SENSING

AI is used to analyze data from sources like satellite imagery and sensors. For example, LSTM networks are trained with different environmental data, including temperature, relative humidity, atmospheric pressure, CO_2 concentration, soil parameters, greenhouse variables, and so on. Applications of AI make the management and calibration of ecosystems easier for assessing forest health, deforestation, environmental restoration, wildlife tracking, habitat assessment, biodiversity analysis, resource conservation, species identification, and natural disaster predictions. For example, AI algorithms analyze camera trap footage, drone imagery, and GPS data to identify and estimate the population size of wildlife, leading to actions in anti-poaching and enhancing protection of diverse species. AI methods are CNNs, random forest, SVMs, RNNs, support vector regression (SVR), and hybrid models.[F17]

6.2.17   FARMING AND AGRICULTURE

Soil monitoring, harvesting, water management, disease protection, and livestock monitoring are growing concerns in the farming industry. All these require cutting-edge technologies, tools, and strategies implemented for optimizing the management. Smart devices, autonomous machinery, IoTs, and wireless sensor networks (WSNs) are a few examples of relatively new methods employed in the agriculture and farming sectors. Modern methods and AI algorithms are helping farmers to improve their output. Some AI methods used in calibrations are LSTMs, CNNs, and many others.[F18] [F19]

6.2.18 FIRE DETECTION

Advancements in deep learning and the IoTs enable early fire detection through vision-based systems, thus reducing ecological, social, and economic damage. AI methods find applications in fire detection, fire response, and fire prevention. Effective deployment of AI-assisted edge devices, drones, firefighting robots, fire suppression systems, and CCTVs plays an important role in optimal performance. A common AI model is the use of CNN algorithms in real-time operation. Other AI methods are random forest, transfer learning, LSTMs, and others.[F20]

6.2.19 HUMAN MACHINE INTERFACE

AI models provide powerful and reliable human–machine interfaces (HMIs). For example, an HMI is an important concept in wearable technologies signifying the human interaction with physical, digital, and mixed environments. Take an armband; it acquires stable biosignals in the form of air pressure in response to muscle activities. Monitored muscle activities consist of contraction and relaxation, causing deformation in the armband's pressure-sensitive chambers. HMI is a highly investigated area due to its applications in industry, driverless vehicles, and so on; hence, a wider and a range of AI models is employed. Natural language processing (NLPs), image recognition systems, and generative AI are some of the methods used in HMI-related applications and calibrations.[F21]

6.2.20 IMAGE QUALITY

Image quality is a broad subject varying from computer-generated images to satellite-based images. In earth observation applications, for instance, data-driven approaches are used to address the calibration challenges for utilizing near-earth hyperspectral data in applications such as agriculture and water resources. A data-driven calibration includes robust algorithms for radiometric calibration (RC), bidirectional reflectance distribution function correction, reflectance normalization, and soil and shadow masking. Generative AIs and diffusion models are known to produce quality images. There are many commercial image calibrators such as Calib, PhotoDirector, MyEdit, Remini, Vectocon, and others.[F22]

6.2.21 INDUSTRY

AI is revolutionizing various industries, including food, drug discovery, e-commerce, chemical industries, cosmetics, tourism, automobile, mechanical, environmental management, gaming, textile, entertainment, and enzymatic design industries. The integration of AI facilitates efficient structuring, administration, design, productivity, manufacturing, promotion, and accessibility of industrial products. All forms of AI models find applications in industrial calibration systems.[F23]

6.2.22 INTERNET OF THINGS

IoTs are a large number of networks of connected devices. By using AI, IoT devices result in enhanced operations by making decisions, automating tasks, and enabling intelligent device-to-device communications without human intervention. There are many applications of IoTs, such as smart homes, manufacturing, healthcare, transportation, agriculture, and retail. The calibration needs of IoT devices depend on the device itself and the nature of the measurand. An example is the sensor anomaly detections. Anomaly detection is necessary to mitigate the impact of faulty sensors on system performance. Another example is the IoT-based incubator analyzer, serving as a measuring tool and calibration tool for diverse parameters such as temperature, mattress temperature, humidity, airflow, and noise in infant incubators. All forms of AI models are used in IoT applications.[F24]

6.2.23 Low-Cost Sensors

Distributed low-cost sensor networks are relatively new tools that are widely deployed in pollution monitoring, atmospheric measurements, environmental monitoring, oceanic studies, and so on. Calibration of low-cost sensors using regulatory-grade monitoring stations is an accepted norm, called Golden Standards (GS). Ground truth measurements of the GS are used to calibrate neighboring nodes, and the information propagates to other sensors in the measurement network. AI techniques like recursive least-squares, LSTMs, CNNs, and others are used in distributed low-cost sensor networks.[F25] [F26]

6.2.24 Magnetometers

Magnetometers are instruments that measure single or triaxial magnetic fields. They are applied in geological surveys, mobile robots, transportation systems, navigation, and submarine detection. AI-based calibration methods of magnetometers reduce errors and improve calibration accuracy. Some of the methods used in calibrations are CNNs, RNNs, Levenberg–Marquardt back propagation NNs (BPNNs), and others.[F27]

6.2.25 Marketing

Various calibration methods are used in marketing. A popular method is the marketing mix models (MMMs), which is a statistical technique integrated with AI models. Calibrations involve fine-tuning the models to ensure that predictions align closely with the incremental changes or estimates. Calibration adjusts the model parameters based on historical data and market experiments. The use of AI in marketing assists in reducing human mistakes, analyzing massive amounts of data, speeding up data processing, and improving stock control. The most AI models used in marketing are predictive and generative AI models and NLP models.[F28]

6.2.26 Medical

AI in medicine is used extensively in medical imaging, analyzing genomic data, surgical procedures, symptom checks, healthcare, diagnosis, treatment, drug discovery, medical record management, and patient communications. An example is cardiac intervention, in which a hybrid CNN-transformer model is used for assessment of angiography-based non-invasive fractional flow-reserve (FFR) and instantaneous wave-free ratio (iFR) of intermediate coronary stenosis. This model predicts if a coronary artery stenosis is hemodynamically significant and provides direct FFR and iFR estimates. A combination of regression and classification branches forces the model to focus on the cut-off region of FFR, which is critical in decision-making. Calibration related to other AI modes are NLPs, self-aware AI, decision trees, and others.[F29] [F30]

6.2.27   MICROELECTROMECHANICAL SYSTEMS (MEMS)

MEMS are cost-effective, self-contained, small-sized, low-power devices used in various sensing applications. Despite their advantages, MEMS-based devices can be noisy, containing systematic and stochastic errors. Though stochastic noises can be characterized and minimized with various analyses and algorithms, systematic errors need to be identified and removed with suitable calibration processes. Typical AI models investigated on MEMS are RNNs, LSTM-RNN, LSTM, SVM, BPNN, GA, Adaboosting, GAN, robust heteroscedastic probabilistic NN, randomized general regression network, and others.[F31] [F32]

6.2.28   ON-DEVICE LEARNING

Due to many advantages, implementation of AI is moving increasingly to end devices such as autonomous vehicles, wearables, smart watches, smart homes, robots, and home appliances. On-device AI models perform regression and inference tasks under extreme memory and processing constraints. To overcome these constraints, many methods have been developed for real-time decision making, reducing latency, and selecting the relevant data on devices. One of the methods is based on Gaussian radial basis function networks. In this method, some hidden neurons are dynamically removed as the need arises. Other methods involve omission of individual features and some data while still being able to correctly predict the result without the need for retraining. On-device learning improves measurement and data flow accuracy. Typical device training and learning methods are TinyML, neural architecture search, neural processing units, and others.[F33] [F34]

6.2.29   PEDESTRIAN DETECTION

Pedestrian detection is identification and tracking of human movements by using sensors and computer vision methods. It finds applications in injury prevention, industrial safety, autonomous vehicles, crowd control, and other areas. Apart from implementation of AI-based computer vision methods, there are numerous models such as feature calibration network, region CNN (R-CNN), mask R-CNNs, fully convolutional one stage, and you only look once (YOLO). Also, commercial pedestrian detection packages are available, like AiVA, Blaxtair Origin, VTPA-30, and others.[F35]

6.2.30 POLLUTION

AI-driven pollution detections use many methods, ranging from deployment of distributed low-cost sensors to satellite imagery. Land-based pollution monitoring systems use wireless networks for data transmission. Various portable pollution monitoring sensors work cooperatively with fixed reference sensors mounted in identified pollution intensive areas. Different pollution sensors may favor different calibration models, and some sensors may favor different AI models. AI models capture non-linear data from individual sensors and integrate it with domain-specific information, meteorology, background pollution, and temporal characteristics. AI methods in pollution detections include linear SVR, CNNs, random forest, SVMs, RNNs, convolutional deep LSTM, and others.[F36]

6.2.31 POSITION

Position sensing is applied across various fields, from health monitoring and manufacturing to the position determination of spaceships. Methods used in position sensing are as diverse as their application areas. In indoor applications, some position and motion tracking systems are based on bluetooth low energy and ultra-wideband wireless networks. Other position sensing systems involve inertial sensors, which are widely utilized in smartphones, drones, vehicles, and wearable devices, playing a crucial role in enabling ubiquitous and reliable localization. Inertial sensor-based positioning sensing is essential in various applications, including personal navigation, location-based security, healthcare, unmanned aerial vehicle (UAV) robots, and human-device interaction. AI methods deployed in position sensing include LSBoost, multiagent deep reinforcement learning, SVMs, CNNs, LSTM, and others.[F37] [F38]

6.2.32 QUANTUM CALIBRATION

Quantum computers offer a new platform in computing. Quantum-related hardware and quantum computing enable larger AI models to run much faster than conventional computing. Calibration of quantum computers is one of the main bottlenecks in scaling quantum systems. Physical qubits must be carefully calibrated since quantum processors are sensitive to the external environment and hardware parameters drifting during operation, thus affecting gate fidelity. Calibration techniques require complex and lengthy measurements to independently control the different parameters of each gate and are unscalable to large quantum systems. AI models such as Bayesian optimization help the design of more functional devices. Some of the AI models tried in quantum calibrations are quantum NLP and quantum RNN and, for NLP tasks, GRU and LSTMs.[F39] [F40]

6.2.33 RADIOMETRIC

Radiometric elements are radioactive isotopes used in radiometric sensing, including uranium, thorium, potassium, and rubidium. They find applications in age determination of fossils and rocks, precise temperature measurements, thermal imaging, geology, paleontology, object classifications, process monitoring, early fire detections, healthcare, agriculture, and many more. RC is a key step in precision sensing. Source-based RC is the most prevailing methodology, where a radiometric source such as a blackbody with known radiation energy is observed by sensors, and the corresponding output is recorded. Artificial intelligence-based calibrations directly generate coefficients, given the state of the sensor parameters. AI methods used in RCs include random forest, SVM, XGBoost, Ross Thick-Li Sparse, maximum diversity GANs, bidirectional feature calibrations, and others.[F41]

6.2.34 Robots

AI-controlled robots are augmented with a variety of devices, sensors, and cameras. They perform complex tasks, operating in diverse environments. There are many different robots, ranging from fixed robots in the manufacturing industry to self-controlled mobile ones. For example, dexterous robots require learning manipulation skills. Deep imitation learning methods, together with transfer learning, give them agility and self-control in performing various tasks. Heterogeneous robots can be grouped to perform collective tasks. A multi-robot system (also called cobots) aims to create robust, flexible, and scalable groups of robots. Two typical algorithms, an ant colony algorithm and a particle swarm algorithm, perform well in distributed multi-robot systems. NLP algorithms help in social learning of robots to perform collaborative tasks and decision-making. Brain-inspired hardware assists in the development of cloud robotics that learn from the experiences of other robots and AI models used in robotic applications are transfer learning, deep imitation learning, reinforcement learning, random forest, RNN, LSTM, CNN, and many others.[F42] [F43] [F44]

6.2.35 Self-Calibration

Self-calibration implies automatic fine-tuning of devices. It also implies self-calibration of AI models such as LLMs. Algorithms such as DRLs can detect sensor faults, such as bias, drift, complete failure, and precision degradation. Such algorithms are also extended for self-calibration faulty sensors automatically using the historical data of the sensor. The self-calibration module can calibrate faulty sensors within seconds, thereby ensuring the uninterrupted availability of accurate monitoring information in many applications such as manufacturing, wearables, and industrial process control. Typical AI algorithms are LSTMs, transfer learning, CNNs, and many others.[F45]

6.2.36 Sensor Fusion

Sensors are the key to the perception of the outside world. Fusion of heterogenous sensors operating in a distributed manner attracts much research and design attention. Using AI models in sensor fusion is becoming an increasingly important technology for many IoT-type applications. Sensor fusion involves a few other technologies like WSNs, Bayesian networks, neuromorphic AI platforms, software-in-the-loop, multi-stream neural networks, RNNs, and transfer learning.[F46]

6.2.37 Soft (Virtual) Sensing

Soft (virtual) sensors refer to mathematical models to estimate phenomena that are difficult to measure. In software nodes, numerous mathematical measurement tasks are performed and processed. ANNs accelerated the development and usage of advanced soft sensors that can identify correlations between measures that could not have been predicted in the past. There are several AI-based soft sensor modules such as Softgen AI, Guru AI, DeepSeek, and others.[F47]

6.2.38 Tactile Sensing

Tactile sensing is the process of detecting contact events such as pressure on the surface. It finds applications in robotics, virtual reality, and prosthetics. Tactile sensing methods are optical capacitive, resistive, and piezoelectric. Electrical resistance types can be used to create large-scale soft tactile sensors that are flexible and robust. Good performance requires a fast and accurate mapping from the sensor's sequential voltage measurements to the distribution of force across its surface. AI-based tactile sensing algorithms are CNN, SVM, KNNs, decision trees, liner discrimination analysis, random forest, RNN, and others.[F48]

6.2.39 Unmanned Vehicles

AI plays a critical role in UAV navigation since it can provide fundamental human control characteristics. There are many different forms of UAVs, ranging from aerial devices (e.g., drones) to unmanned underwater vehicles. Different applications have adopted different AI approaches to make autonomous UAV navigation more efficient. Typical examples of AI models in UAVs are GANs, VAEs, cGANs (conditional GANs), CycleGANs, CNNs, YOLO, LSTMs, Naïve bias, and others.[F49]

6.2.40 Wireless Communications

Machine learning methods in wireless communication systems use deep learning techniques to train massive amounts of data. AI models are applied in different areas of communication systems for signal processing, optimization of operations, noise reduction, antenna location, network planning, coverage estimation, traffic prediction, routing, network automation, and WSN management. Bayesian learning is an example of an AI model used in noise and interference-prone environments. AI methods find wide applications in communication systems and wireless networks, examples are NLP, CNNs, decision trees, and LSTMs.[F50] [F51] [F52] [F53]

6.3 CONCLUSIONS

Calibration of devices and processes provides consistency in readings and reduces errors. Calibration is an extremely well-developed science and technology, as it finds applications in all types of measurements and measurement systems. Using AI models in calibration opens new frontiers. It provides individual device calibration as well as complete system calibration. This chapter highlights that AI models in calibration range from large IoT systems to education and economics.

REFERENCES

F1. A. Roitberg et al., "Is My Driver Observation Model Overconfident? Input-Guided Calibration Networks for Reliable and Interpretable Confidence Estimates," *IEEE Transactions on Intelligent Transportation Systems*, vol. 23, no. 12, pp. 25271–25286, Dec. 2022.

F2. M. A. Zaidan et al., "Intelligent Air Pollution Sensors Calibration for Extreme Events and Drifts Monitoring," *IEEE Transactions on Industrial Informatics*, vol. 19, no. 2, pp. 1366–1379, Feb. 2023.

F3. J. Buelvas, D. Munera, D. P. Tobon, V. J. Aguirre and N. Gaviria, "Data Quality in IoT-Based Air Quality Monitoring Systems: a Systematic Mapping Study," *Water Air Soil Pollution*, vol. 234, p. 248, 2023.

F4. N. Sarker, P. Podder, M. R. H. Mondal, S. S. Shafin and J. Kamruzzaman, "Applications of Machine Learning and Deep Learning in Antenna Design, Optimization, and Selection: A Review," *IEEE Access*, vol. 11, pp. 103890–103915, 2023.

F5. M. Sadiq, N. Sulaiman, M. Mohd Isa, M. N. Hamidon, "A Review on Machine Learning in Smart Antenna: Methods and Techniques," *TEM Journal*, vol. 11, no. 2, pp. 695–705, May 2022.

F6. P. Wei et al., "CROON: Automatic Multi-LiDAR Calibration and Refinement Method in Road Scene," *2022 IEEE/RSJ International Conference on Intelligent Robots and Systems (IROS)*, Kyoto, Japan, pp. 12857–12863, 2022.

F7. J. Guan, L. Pan, C. Wang, S. Yu, L. Gao and X. Zheng, "Trustworthy Sensor Fusion against Inaudible Command Attacks in Advanced Driver-Assistance Systems," *IEEE Internet of Things Journal*, vol. 10, no. 19, pp. 17254–17264, Oct. 2023.

F8. J. Leitner, P. -H. Chiang and S. Dey, "Personalized Blood Pressure Estimation Using Photoplethysmography: A Transfer Learning Approach," *IEEE Journal of Biomedical and Health Informatics*, vol. 26, no. 1, pp. 218–228, Jan. 2022.

F9. J. Fu, L. Lin and Q. Li, "Self-Calibration for Star Sensors," *Sensors*, vol. 24, p. 3698, 2024.

F10. D. Macii, "Basics of Industrial Metrology," *IEEE Instrumentation & Measurement Magazine*, vol. 26, no. 6, pp. 5–12, Sept. 2023.

F11. X. Wang, P. Jia, X. Shen and H. V. Poor, "Intelligent and Low Overhead Network Synchronization for Large-Scale Industrial IoT Systems in the 6G Era," *IEEE Network*, vol. 37, no. 3, pp. 76–84, May/June 2023.

F12. H. Zhai, X. Song, X. Wang and G. Liu, "Design of a Flow Automatic Calibration System Based on the Master Meter and Dynamic Weighing Methods," *IEEE Access*, vol. 12, pp. 37141–37151, 2024.

F13. R. Wu, P. Zhang, N. Xu, X. Hu, L. Chen, L. Zhang and Z. Yang, "FY-3D MERSI On-Orbit Radiometric Calibration from the Lunar View," *Sensors*, vol. 20, no. 4690, pp. 1–13, 2020.

F14. C. C. Tossell, N. L. Tenhundfeld, A. Momen, K. Cooley and E. J. de Visser, "Student Perceptions of ChatGPT Use in a College Essay Assignment: Implications for Learning, Grading, and Trust in Artificial Intelligence," *IEEE Transactions on Learning Technologies*, vol. 17, pp. 1069–1081, 2024.

F15. S. W. Hsiao and C. Y. Sun, "LoRA-like Calibration for Multimodal Deception Detection Using ATSFace Data," *2023 IEEE International Conference on Big Data (BigData)*, Sorrento, Italy, pp. 2163–2172, 2023.

F16. R. R. Kumar et al., "Induction Machine Stator Fault Tracking Using the Growing Curvilinear Component Analysis," *IEEE Access*, vol. 9, pp. 2201–2212, 2021.

F17. O. N. Chisom, P. W. Biu, A. A. Umoh and B. Obaedo, "Reviewing the Role of AI in Environmental Monitoring and Conservation: A Data-Driven Revolution for Our Planet," *World Journal of Advanced Research and Reviews*, vol. 21, no. 1, pp. 161–171, Jan. 2024.

F18. R. J. Martin et al., "XAI-Powered Smart Agriculture Framework for Enhancing Food Productivity and Sustainability," *IEEE Access*, vol. 12, pp. 168412–168427, 2024.

F19. S. I. Hassan, M. M. Alam, U. Illahi, M. A. Al Ghamdi, S. H. Almotiri and M. M. Su'ud, "A Systematic Review on Monitoring and Advanced Control Strategies in Smart Agriculture," *IEEE Access*, vol. 9, pp. 32517–32548, 2021.

F20. N. Dilshad, S. U. Khan, N. S. Alghamdi, T. Taleb and J. Song, "Toward Efficient Fire Detection in IoT Environment: A Modified Attention Network and Large-Scale Data Set," *IEEE Internet of Things Journal*, vol. 11, no. 8, pp. 13467–13481, April 2024.

F21. H. Zhou, C. Tawk and G. Alici, "A Multipurpose Human–Machine Interface via 3D-Printed Pressure-Based Force Myography," *IEEE Transactions on Industrial Informatics*, vol. 20, no. 6, pp. 8838–8849, June 2024.

F22. Vasit Sagan, M. Maimaitijiang, *et al.*, "Data-Driven Artificial Intelligence for Calibration of Hyperspectral Big Data," *IEEE Transactions on Geoscience and Remote Sensing*, vol. 60, pp. 1–20, Article no. 5510320, 2022.

F23. S. Malik, K. Muhammad and Y. Waheed, "Artificial Intelligence and Industrial Applications—A Revolution in Modern Industries," *Ain Shams Engineering Journal*, vol. 15, no. 9, pp. 1–11, Sept. 2024. Open access. https://www.sciencedirect.com/science/article/pii/S2090447924002612 (Accessed on 24 July 2025)

F24. S. K. Maulani, S. Syaifudin, A. M. Maghfiroh and L. F. Wakidi, "Design of Incu Analyzer for IoT-Based Baby Incubator Calibration," *International: Rapid Review: Open Access Journal*, vol. 17, no. 1, pp. 20–28, March 2024.

F25. R. Krüger, P. Karrasch and A. Eltner, "Calibrating Low-Cost Rain Gauge Sensors for Their Applications in Internet of Things (IoT) Infrastructures to Densify Environmental Monitoring Networks," *Geoscientific Instrumentation, Methods and Data Systems*, vol. 13, pp. 163–176, 2024.

F26. T. Becnel, T. Sayahi, K. Kelly and P. E. Gaillardon, "A Recursive Approach to Partially Blind Calibration of a Pollution Sensor Network," *2019 IEEE International Conference on Embedded Software and Systems (ICESS)*, Las Vegas, NV, USA, pp. 1–8, 2019.

F27. J. H. Hong, D. Kang and I. -J. Kim, "A Unified Method for Robust Self-Calibration of 3-D Field Sensor Arrays," *IEEE Transactions on Instrumentation and Measurement*, vol. 70, pp. 1–11, Article no. 1007211, 2021.

F28. S. Verma, R. Sharma, S. Deb and D. Maitra, "Artificial Intelligence in Marketing: Systematic Review and Future Research Direction," *International Journal of Information Management Data Insights,* vol. 1, no. 1, p. 1000002, April 2021.

F29. H. H. Nguyen, M. B. Blaschko, S. Saarakkala and A. Tiulpin, "Clinically Inspired Multi-Agent Transformers for Disease Trajectory Forecasting from Multimodal Data," *IEEE Transactions on Medical Imaging*, vol. 43, no. 1, pp. 529–541, Jan. 2024.

F30. R. Mineo et al., "A Convolutional-Transformer Model for FFR and iFR Assessment from Coronary Angiography," *IEEE Transactions on Medical Imaging*, vol. 43, no. 8, pp. 2866–2877, Aug. 2024.

F31. A. Harindranath and M. Arora, "A Systematic Review of User-Conducted Calibration Methods for MEMS-Based IMUs," *Measurement*, vol. 225, 114001, Feb. 2024.

F32. A. L. Shestakov, "Dynamic Measuring Methods: A Review," *Acta IMEKO*, vol. 8, no. 1, pp. 64–76, March 2019.

F33. K. Sunaga, M. Kondo and H. Matsutani, "Addressing the Gap Between Training Data and Deployed Environment by On-Device Learning," *IEEE Micro*, vol. 43, no. 6, pp. 66–73, Nov.-Dec. 2023.

F34. F. Saccani, D. Pau and M. Amoretti, "Learning Pressure Sensor Drifts from Zero Deployability Budget," *IEEE Sensors Letters*, vol. 8, no. 8, pp. 1–4, Aug. 2024,

F35. T. Zhang, Q. Ye, B. Zhang, J. Liu, X. Zhang and Q. Tian, "Feature Calibration Network for Occluded Pedestrian Detection," *IEEE Transactions on Intelligent Transportation Systems*, vol. 23, no. 5, pp. 4151–4163, May 2022.

F36. Y. Han, S. Song, Y. Yu, J. C. K. Lam and V. O. K. Li, "UNI-CAL: A Universal AI-Driven Model for Air Pollutant Sensor Calibration with Domain-Specific Knowledge Inputs," *IEEE Access,* vol. XX, pp. 1–14, 2024.

F37. C. Chen and X. Pan, "Deep Learning for Inertial Positioning: A Survey," *IEEE Transactions on Intelligent Transportation Systems*, vol. 25, no. 9, pp. 10506–10523, Sept. 2024.

F38. X. Li et al., "A Cooperative Relative Localization System for Distributed Multi-Agent Networks," *IEEE Transactions on Vehicular Technology*, vol. 72, no. 11, pp. 14828–14843, Nov. 2023.

F39. K. Gautam and C. W. Ahn, "Quantum Path Integral Approach for Vehicle Routing Optimization with Limited Qubit," *IEEE Transactions on Intelligent Transportation Systems*, vol. 25, no. 5, pp. 3244–3258, May 2024.

F40. O. Shindi, Q. Yu, P. Girdhar and D. Dong, "Model-Free Quantum Gate Design and Calibration Using Deep Reinforcement Learning," *IEEE Transactions on Artificial Intelligence*, vol. 5, no. 1, pp. 346–357, Jan. 2024.

F41. B. Chen, et al., "Towards Artificial Intelligence Based Radiometric Calibration for In-Orbit Remote Sensing Satellites," *Research Square*, pp. 1–26, 2024. Open Source https://www.researchsquare.com/article/rs-3826695/v1 (Accessed on 24 July 2025).

F42. H. Karim, D. Gupta and S. Sitharaman, "Securing LLM Workloads with NIST AI RMF in the Internet of Robotic Things," *IEEE Access*, vol. 13, pp. 69631–69649, 2025.

F43. Z. Nie, K. -C. Chen and K. J. Kim, "Social-Learning Coordination of Collaborative Multi-Robot Systems Achieves Resilient Production in a Smart Factory," *IEEE Transactions on Automation Science and Engineering*, vol. 22, pp. 6009–6023, 2025.

F44. S. Qiu, M. Wang and M. R. Kermani, "A Modern Solution for an Old Calibration Problem," *IEEE Instrumentation & Measurement Magazine*, vol. 24, no. 3, pp. 28–35, May 2021.

F45. X. Cao, J. W. Crandall and M. A. Goodrich, "Improving Robot Proficiency Self-Assessment via Meta-Assessment," *IEEE Robotics and Automation Letters*, vol. 8, no. 11, pp. 7297–7303, Nov. 2023.

F46. F. Zijie et al., "Wireless Sensor Networks in the Internet of Things: Review, Techniques, Challenges, and Future Directions," *Indonesian Journal of Electrical Engineering and Computer Science*, vol. 31, no. 2, pp. 1190–1200, Aug. 2023.

F47. J. Miettinen, T. Tiainen, R. Viitala, K. Hiekkanen and R. Viitala, "Bidirectional LSTM-Based Soft Sensor for Rotor Displacement Trajectory Estimation," *IEEE Access*, vol. 9, pp. 167556–167569, 2021.

F48. H. Lee et al., "Predicting the Force Map of an ERT-Based Tactile Sensor Using Simulation and Deep Networks," *IEEE Transactions on Automation Science and Engineering*, vol. 20, no. 1, pp. 425–439, Jan. 2023.

F49. S. Rezwan and W. Choi, "Artificial Intelligence Approaches for UAV Navigation: Recent Advances and Future Challenges," *IEEE Access*, vol. 10, pp. 26320–26339, 2022.

F50. M. Zecchin, S. Park, O. Simeone, M. Kountouris and D. Gesbert, "Robust Bayesian Learning for Reliable Wireless AI: Framework and Applications," *IEEE Transactions on Cognitive Communications and Networking*, vol. 9, no. 4, pp. 897–912, Aug. 2023.

F51. H. N. Qureshi, A. Imran and A. Abu-Dayya, "Enhanced MDT-Based Performance Estimation for AI Driven Optimization in Future Cellular Networks," *IEEE Access*, vol. 8, pp. 161406–161426, 2020.

F52. D. F. S. Fernandes, A. Raimundo, F. Cercas, P. J. A. Sebastião, R. Dinis and L. S. Ferreira, "Comparison of Artificial Intelligence and Semi-Empirical Methodologies for Estimation of Coverage in Mobile Networks," *IEEE Access*, vol. 8, pp. 139803–139812, 2020.

F53. Y. Guo, B. Yuan, A. Su, C. Shao and Y. Gao, "Calibration for Improving the Medium-Range Soil Temperature Forecast of a Semiarid Region over Tibet: A Case Study," *Atmosphere*, 15, 591, pp. 2–32, 2024.

List of Abbreviations

AC	Actor Critic
ACO	Ant Colony Optimization
ADC	Analog to Digital Converter
AE	Autoencoder
AE-DNN	Autoencoder DNN
AI	Artificial Intelligence
AIMC	Analog In-Memory Computing
ANN	Artificial Neural Network
ANOVA	Analysis of Variance
ANSI	American National Standards Institute
APLAC	Asia Pacific Laboratory Accreditation Cooperation
ARNN	Autoregressive NN
ASIC	Application-Specific Integrated Chips
BCS	British Calibration Services
BFC	Bidirectional Feature Calibration
BiGAN	Bidirectional GAN
BiLSTM	Bidirectional Long Short-Term Memory
BIPM	Bureau International des Poids et Mesures
BIRC	Balance Iterative Reduction Clustering
BLR	Bayesian Linear Regression
BN	Bayesian Network
BNN	Bayesian Neural Networks
BPNN	Back Propagation Neural Network
cGAN	Conditional GANs
CBC	Centroid-Based Clustering Or Constraint-Based Clustering
CCO	Current-Controlled Oscillator
CCSD	Calibration Check Standard Database
CDLSTM	Convolutional Deep Long Short-Term Memory
CDMA	Code Division Multiple Access
CF-MNN	Cascade-Forward MNN
CGPM	Conférence Générale des Poids et Mesures
CI	Computational Intelligence
CLAC	Lyapunov-Based Actor Critic
CLDNN	Convolution Long Short-Term Deep NN
CMOS	Complementary Metal Oxide Semiconductors
CNN	Convolutional Neural Network
CPAI	Cyber-Physical AI
CPU	Central Processing Units
CQI	Continuous Quality Improvement
DAC	Digital-to-Analog Converter
DAQ	Data Acquisition
DARPA	Defense Advanced Research Projects Agency

DBC	Density-Based Clustering
DBSCAN	Density-Based Spatial Clustering of Applications with Noise
DL	Deep Learning
DNN	Deep Neural Network
DP	Dynamic Programming
DQN	Deep Q-Networks
DQD	Data Quality Directive
DQO	Data Quality Objective
DRL	Deep Reinforcement Learning
DSSS	Deep Space Surveillance Systems
DSN	Distributed Sensor Networks
DT	Decision Tree
DTR	Decision Tree Regression
DUNE	Deep Underground Neutrino Experiment
DUT	Device Under Test
EC	Edge Computing
EEoI	Early Exit of Inference
EKF	Extended Kalman Filters
ELM	Extreme Learning Machine
EM	Expectation Maximization
EN	Elastic Net
EQA	External Quality Assessment
EOS	Earth Observing System
FC	Fuzzy Clustering
FCC	Fuzzy C-Mean Clustering
FC-Net	Feature Calibration Network
FCOS	Fully Convolutional One Stage
FFBPNN	Feedforward Back Propagation Neural Network
FFR	Fractional Flow-Reserve
FL	Federated Learning
FNN	Feedforward Neural Network
FPGA	Field-Programmable Gate Array
FTL	Federated Transfer Learning
GA	Genetic Algorithm
GAI	Generative AI
GAN	Generative Adversarial Networks
GBM	Gradient Boosting Machine
GDAS	Global Data Assimilation System
GDP	Gross Domestic Product
GMM	Gaussian Mixture Model
GNB	Gaussian Naïve Bayes
GOF	Goodness of Fit
GPS	Global Positioning System
GPT	Generative Pretrained Transformer
GPU	Graphic Processing Unit
GRBF	Gaussian Radial Basis Function

GRU	Gated Recurrent Unit
GSA	Global Search Algorithm
GSO	Geostationary Satellite Orbits
GUM	Guided to Uncertainty Measurement
HBC	Hierarchical-Based Clustering
HDC	Hyperdimensional Computing
HITL	Human-in-the Loop
HMI	Human–Machine Interfaces
HTTP	Hypertext Transfer Protocol
iCAL	Internet Calibration
iFR	Instantaneous wave-Free Ratio
IAAC	Inter-American Accreditation Cooperation
IAF	International Accreditation Forum
IC	Integrated Circuit
IEC	International Electrotechnical Commission
IEEE	Institute of Electrical and Electronics Engineers
ILAC	International Laboratory Accreditation Cooperation
I/O	Input/Output
IoT	Internet of Things
IQC	Internal Quality Checks
ISO	International Organization for Standardization
ISSC	Information System to Support Calibrations
KAA	Kernel Approximation Algorithm
KD	Knowledge Distillation
KF	Kalman Filter
KMC	K-Means Clustering
KNN	K-Nearest Neighbor
LLM	Large Language Model
LM	Language Model
LMBP	Levenberg Marquardt Back Propagation
LMBP-ANN	Levenberg Marquardt Back Propagation Artificial Neural Network
LMMSE	Linear Minimum Mean Square Error
LOF	Lack of Fit
LR	Linear Regression
LSTM	Long Short-Term Memory
MADRL	Multiagent Deep Reinforcement Learning
MAE	Mean Absolute Error
MC	Monte Carlo
MCU	Microcontroller Units
MDGAN	Maximum Diversity GAN
MDP	Markov Decision Process
MIIDAPS	Multi-Instrument Inversion and Data Assimilation Preprocessing System
MIL-STD	Military Standard
ML	Machine Learning
MLP	Multi-Layer Perceptron

MLR	Multiple Linear Regression
MMM	Marketing Mix Models
MMQT	MultiModal QuantTree
MMTM	Multimodal Transfer Module
MNN	Multi-Layer Neural Networks
MPS	Model Predictive Control
MRB	Multi-Armed Bandit
MSE	Mean Square Error
NAAU	National Accreditation Agency of Ukraine
NAS	Neural Architecture Search
NASA	National Aeronautics and Space Administration
NATA	National Association of Testing Authorities
NB	Naïve Bayes
NBR	Negative Binomial Regression
NIST	National Institute of Standards and Technology
NLP	Natural Language Processing
NMSE	Normalized Mean Square Error
NN	Neural Network
NPL	National Physical Laboratory
NPU	Neural Processing Unit
OEM	Original Equipment Manufacturer
OR	Ordinal Regression
ODE	Ordinary Differential Equations
PAI	Predictive AI
PBC	Partition-Based Clustering PBC
PCA	Principal Component Analysis
PCR	Principal Component Regression
PDE	Partial Differential Equations
PDS	Photon Detection System
PIDL	Physics-Informed Deep Learning
PIML	Physically Informed Machine Learning
PINN	Physic-Informed Neural Network
PLS	Partial Least Squares
PLS-DA	Partial Least Squares Discriminant Analysis
PMP	Pontryagin's Minimum Principle
PMT	Photomultiplier Tube
PPG	Photoplethysmogram
PR	Poisson Regression
PRL	Parallel Reinforcement Learning
PSO	Particle Swarm Optimization
QNLP	Quantum NLP
QoD	Quality of Data
QR	Quantile Regression
QRNN	Quantum RNN
R^2	Coefficient of Determination
R-CNN	Region-CNN

RCC	Reliance Calibration Cues
RESD	Residual Standard Deviation
RF	Random Forest
RFR	Random Forest Regression
RGRN	Randomized General Regression Network
RHPNN	Robust Heteroscedastic Probabilistic NN
RL	Reinforcement Learning
RLHF	Reinforcement Learning with Human Feedback
RMSE	Root Mean Square Error
RNN	Recurrent Neural Network
RR	Ridge Regression
SA	Simulated Annealing
SAC	Soft Actor Critic
SCC	Standards Council of Canada
SCE	Shuffled Complex Evaluation
SDP	Stochastic Dynamic Programming
SGD	Stochastic Gradient Descent
SI	Systeme International d'Unites
SIL	Software-in-the-Loop
SL	Supervised Learning
SNN	Spiking Neuron Network
SoC	System-on-Chip
SR	Stepwise Regression
SSIM	Smart Sensors and Integrated Microsystem
SSL	Semi-Supervised Learning
SVM	Support Vector Machine
SVR	Support Vector Regression
TCP/IP	Transmission Control Protocol/Internet Protocol
TED	Transducer Electronic Datasheets
TL	Transfer Learning
TUR	Test Uncertainty Ratio
UAV	Unmanned Aerial Vehicle
UKF	Unscented Kalman Filters
UWB	Ultra-Wideband
VAE	Variational Autoencoder
VCO	Voltage-Controlled Oscillator
VIM	Vocabulary in Metrology
WHO	World Health Organization
WSN	Wireless Sensor Network
XAI	Explainable Artificial Intelligence
YOLO	You Only Look Once
$\Sigma\Delta$M	Delta-Sigma Modulation

Index

Note: **Bold** page numbers refer to tables and *italic* page numbers refer to figures.

For Product Safety Concerns and Information please contact our EU
representative GPSR@taylorandfrancis.com
Taylor & Francis Verlag GmbH, Kaufingerstraße 24, 80331 München, Germany

www.ingramcontent.com/pod-product-compliance
Lightning Source LLC
Chambersburg PA
CBHW031953180326
41458CB00006B/1706

9 781032 968070